Evaluation of sensor-based precision methods for mechanical weed control in arable crops

Evaluation of sensor-based precision methods for mechanical weed control in arable crops

Dissertation to obtain the doctoral degree of Agricultural Sciences (Dr. sc. agr.)

Faculty of Agricultural Sciences
University of Hohenheim

Institute of Phytomedicine (360)
Department of Weed Science (360b)
Prof. Dr. Roland Gerhards

submitted by

Jannis Machleb

from Hildburghausen

Stuttgart-Hohenheim
August 2020

Bibliografische Information der Deutschen Nationalbibliothek
Die Deutsche Nationalbibliothek verzeichnet diese Publikation in der Deutschen Nationalbibliografie; detaillierte bibliographische Daten sind im Internet über http://dnb.d-nb.de abrufbar.
1. Aufl. - Göttingen: Cuvillier, 2020
Zugl.: Hohenheim, Univ., Diss., 2020

D100

Nonnenstieg 8, 37075 Göttingen
Telefon: 0551-54724-0
Telefax: 0551-54724-21
www.cuvillier.de

1. Auflage, 2020
Gedruckt auf umweltfreundlichem, säurefreiem Papier aus nachhaltiger Forstwirtschaft.

ISBN 978-3-7369-7248-3
eISBN 978-3-7369-6248-4

TABLE OF CONTENTS

1 General Introduction 11

1.1 Objectives of the Thesis 13

1.2 Structure of the Thesis 14

2 Publications 15

2.1 Sensor-based mechanical weed control: Present state and prospects 16

2.1.1 Abstract 16

2.1.2 Introduction 18

2.1.3 Autonomous Robots 22

2.1.4 Half Autonomous Tractor-pulled Implements 28

2.1.4.1 IIa Machine Vision Systems 28

2.1.4.2 IIb GNSS Sensor Systems 31

2.1.4.3 IIc Laser and Ultrasonic Sensor Systems 32

2.1.5 Discussion and Outlook 33

2.1.6 Conclusion 42

2.1.7 Declaration of Competing Interest 43

2.1.8 References 44

2.2 Adjustment of weed hoeing to narrowly spaced cereals 53

2.2.1 Abstract 53

2.2.2 Introduction 55

2.2.3 Materials and Methods 56

2.2.3.1 Experimental Sites and Design 56

2.2.3.2 Weeding tool description and implementation 57

2.2.3.2.1 Goosefoot Sweeps 59

2.2.3.2.2 No-Till Sweeps 60

2.2.3.2.3 Down-Cut Side Knives 61

2.2.3.3 Data Collection 62

2.2.3.4 Data Analysis 63

2.2.4 Results 64

2.2.4.1 Results at Ihinger Hof 64

2.2.4.2 Results at Kleinhohenheim 66

2.2.5 Discussion 68

2.2.6 Conclusions 72

2.2.7 Appendix A 74

2.2.8 References 76

2.3 Sensor-based intrarow mechanical weed control in sugar beets with motorized finger weeders 79

2.3.1 Summary 79

2.3.2 Introduction 81

2.3.3 Materials and Methods 83

2.3.3.1 Overview and Experimental Site 83

2.3.3.2 Herbicides and application details 84

2.3.3.3 General set-up of the hoe 85

2.3.3.4 Set-up of the motorized finger weeders 87

2.3.3.5 Implementation of the mechanical treatments 89

2.3.3.6 Description of the sensor system 90

2.3.3.7 Data acquisition 93

2.3.3.8 Data analysis 94

2.3.4 Results 95

2.3.4.1 Weed Density and Weed Control Efficacy 95

2.3.4.2 Sugar beet yield 97

2.3.5 Discussion 98

2.3.6 Acknowledgements 101

2.3.7 Conflict of Interest 101

2.3.8 References 102

3 General Discussion 105

4 Summary 119

5 Zusammenfassung 122

6 General References 126

7 Eidesstattliche Versicherung 134

8 DANKSAGUNG 135

9 CURRICULUM VITAE 136

LIST OF FIGURES

Figure 2.1.2-1: Harrowing peas (A) and hoeing in soybean (B). 20

Figure 2.1.5-1: Depiction of the three weeding zones: A = inter-row space, B = intrarow space, C = close-to-crop space. Weeds are not labeled. Crop plants are labeled as "Crop". Adopted and modified from (Pérez-Ruiz *et al.*, 2012). 35

Figure 2.1.5-2: A camera-steered hoe follows curved rows of summer barley 37

Figure 2.2.3-1: Overview of the parallelogram system: two K.U.L.T. DUO-parallelograms with no-till sweeps for 150 mm row spacing 58

Figure 2.2.3-2: Group image of the three cultivation sweeps used in this study. From left to right: goosefoot sweep, no-till sweep and down-cut side knife. The depicted sweeps were adjusted to a row spacing of 150 mm. 59

Figure 2.2.4-1: (a) Spring barley grain yield; (b) spring barley dry mass recorded at Ihinger Hof in 2017. Means with the same letter are not significantly different according to Duncan's multiple range test at $\alpha \leq 0.05$. CON = untreated control, HERB = herbicide application, GFS(4) = goosefoot sweeps 4 km h^{-1}, GFS(6) = goosefoot sweeps 6 km h^{-1}, NTS(4) = no-till sweeps 4 km h^{-1}, NTS(6) = no-till sweeps 6 km h^{-1}, DSK = down-cut side knife 4 km h^{-1}. 65

Figure 2.2.4-2: (a) Mean weed control efficacy in spring barley; (b) weed dry mass in spring barley recorded at Ihinger Hof in 2017. Means with the same letter are not significantly different according to Duncan's multiple range test at $\alpha \leq 0.05$. CON = untreated control, HERB = herbicide application, GFS(4) = goosefoot sweeps 4 km h^{-1}, GFS(6) = goosefoot sweeps 6 km h^{-1}, NTS(4) = no-till sweeps 4 km h^{-1}, NTS(6) = no-till sweeps 6 km h^{-1}, DSK = down-cut side knife 4 km h^{-1}. 66

Figure 2.2.4-3: (a) Spring oats grain yield; (b) spring oats dry mass recorded at Kleinhohenheim in 2017. Means with the same letter are not statistically different according to Duncan's multiple range test at $\alpha \leq 0.05$. CON = untreated control, MANW = manual weeding, GFS(4) = goosefoot sweeps 4 km h^{-1}, GFS(6) = goosefoot sweeps 6 km h^{-1}, NTS(4) = no-till sweeps 4 km h^{-1}, NTS(6) = no-till sweeps 6 km h^{-1}, DSK = down-cut side knife 4 km h^{-1}. 67

Figure 2.2.4-4: (a) Mean weed control efficacy in spring oats; (b) weed dry mass in spring oats recorded at Kleinhohenheim in 2017. Means with the same letter are not statistically different according to Duncan's multiple range test at $\alpha \leq 0.05$. CON = untreated control, MANW = manual weeding, GFS(4) = goosefoot sweeps 4 km h^{-1}, GFS(6) = goosefoot sweeps 6 km h^{-1}, NTS(4) = no-till sweeps 4 km h^{-1}, NTS(6) = no-till sweeps 6 km h^{-1}, DSK = down-cut side knife 4 km h^{-1}. 68

Figure 2.3.3-1: The set-up of the hoe used in the experiments at Ihinger Hof in 2017 and 2018. The image displays one plot width (six sugar beet rows). 1 = Garford Robocrop, 2 = Garford Robocrop camera for interrow weeding, 3 = Argus hoe frame, 4 = parallelograms with goosefoot sweeps, 5 = conventional finger weeders, 6 = bi-spectral camera for intrarow weeding (colour blue), 7 = odometry wheel, 8 = motorized finger weeders (colour blue), 9 & 10 = the two centre sugar beet rows of each plot that were treated with the MFW and used for harvesting. Design ele-ments by K.U.L.T. Germany 86

Figure 2.3.3-2: Diagram showing the rotation direction and speed of the motorized finger weeders. 87

Figure 2.3.3-3: The motorized finger weeders (front) and conventional finger weeders (back). 88

Figure 2.3.3-4: Flowchart describing the sensor system. 92

Figure 2.3.3-5: The bi-spectral camera for intrarow weeding: the lens is surrounded by the infrared LEDs. 92

Figure 2.3.3-6: The frame (0.5 x 0.5 m) that was used for the measurements of the weed density in 2017 and 2018. 93

Figure 2.3.4-1: The bars represent the mean sugar beet yield (t ha^{-1}) recorded for each treatment in 2017 and 2018. Different letters above a bar and within the same graph (2017 or 2018) indicate significant differences between the treatments according to the Tukey HSD-Test at $\alpha \leq 0.05$. 97

LIST OF TABLES

Table 2.1.2-1: Basic requirements for efficient sensor-guided mechanical weeding machines. 21

Table 2.1.3-1: Robots in agriculture and other environments, adapted and modified from Bechar and Vigneault (2017). 22

Table 2.1.5-1: Prospects and future challenges of sensor-based mechanical weed control. 34

Table 2.1.5-2: Overview of commercial sensor-based mechanical weed control systems in agriculture. 36

Table 2.1.5-3: Comparison of GNSS and machine vision guidance accuracy in different studies. 38

Table 2.2.3-1: Top view of the commercial and adjusted goosefoot sweeps with the narrow row widths of 150 and 125 mm at Ihinger Hof and Kleinhohenheim, respectively. Hatched lines symbolize the area cut off from the commercial 160 mm wide goosefoot sweep. The plants to the left and right of the sweeps symbolize cereal rows. 60

Table 2.2.3-2: Top view of the commercial and the adjusted no-till sweeps to the narrow row widths of 150 and 125 mm at Ihinger Hof and Kleinhohenheim, respectively. Hatched lines symbolize the area cut off from the commercial 160 mm wide no-till sweep. The plants left and right of the sweeps symbolize cereal rows 61

Table 2.2.3-3: Side view of the commercial down-cut side knife and top view of the down-cut side knives adjusted to the narrow row widths of 150 and 125 mm at Ihinger Hof and Kleinhohenheim, respectively. Hatched lines (here only shown for the 125 mm row width) indicate the blade length that was cut off from the commercial knife blade to fit the side knives between the cereal rows. The plants left and right of the side knives symbolize cereal rows. 62

Table 2.2.7-1: Overview of the treatments at Ihinger Hof and Kleinhohenheim in 2017. 74

Table 2.2.7-2: Weed composition at the two research locations Ihinger Hof and Kleinhohenheim in 2017 75

Table 2.3.3-1: Description of the treatments at Ihinger Hof in 2017 and 2018. 83

Table 2.3.3-2: Herbicide type and application time (BBCH of the sugar beets) at Ihinger Hof in 2017 and 2018 84

Table 2.3.3-3: Time of application of the mechanical treatments at Ihinger Hof according to the number of developed leaves and the BBCH stage (Lancashire *et al.*, 1991) 89

Table 2.3.4-1: The results obtained for the mean weed density (plants m^{-2}) of the intrarow and inter-row area measured three days after the final application of each treatment in 2017 and 2018. Additionally, the results for the intrarow weed control efficacy (%) are shown. Means with different letters within the same column indicate significant differences between the treatments according to the Tukey HSD-Test at $\alpha \leq 0.05$. WCE = weed control efficacy 96

Table 2.3.4-2: The number of sugar beets per hectare before the application of the treatments (crop emergence) and at the harvest date in 2017 and 2018. 98

1 GENERAL INTRODUCTION

Worldwide crop production is constantly threatened by abiotic and biotic yield reducing factors. Abiotic factors include water, temperature, and the competition for space as well as inorganic nutrients whereas biotic factors are weeds, animal pests and pathogens. Potential average yield losses without weed control alone can be as high as 23% (wheat), 40.3% (maize) and 37% (soybean) (Oerke, 2006). Some crops such as sugar beets are extremely sensitive to weed competition and can result in yield losses of up to 95% (Petersen, 2004). Therefore, in a lot of cases weeds are the major cause for yield losses (Oerke, 2006). Along with the competition for resources, weeds can complicate the harvest and provide shelter for insect pests and pathogens (Holzner and Numata, 1982; Wisler and Norris, 2005; Mohler, 2007; Savary *et al.*, 2012; Vincent, Panneton and Fleurat-Lessard, 2013). Therefore, effective weed control is an essential part of plant protection.

The use of synthetic herbicides revolutionized agriculture and it has been the main strategy to control weeds in arable crops in developed countries since the 1950s (Abbas *et al.*, 2018). Prior to the introduction of modern herbicides, labor intensive manual or mechanical weeding were the most common form of weed control. Applying herbicides reduced the need for human labor while simultaneously improving crop quality and quantity at lower costs (Zimdahl, 2015, p. 104). Herbicide use also helps to conserve natural resources including water, soil and energy where reduced tillage is practiced (Gianessi, 2013). However, concerns are growing over the possible adverse effects of herbicides on public health, the environment and non-target organisms (Özkara, Akyil and Konuk, 2016). An indirect consequence of herbicide use is the elimination of possible insect food sources by reducing weed diversity or by decreasing the amount of flowers and seeds produced by weed plants growing in field margins (Schmitz, Schäfer and Brühl, 2014; Hahn *et al.*, 2015; Kovács-Hostyánszki *et al.*, 2017). Armengot *et al.* (2013) showed that mechanical weed control with a harrow did not negatively impact weed diversity in winter cereals while maintaining the same yield levels as the weed-free control plots. The increased occurrence of herbicide resistant weeds during the past years is another major factor that triggered a rethinking process for plant protection practices with herbicides (Moss, 1993; Buhler, Liebman and Obrycki, 2006; Reddy and Norsworthy, 2010; Délye, Jasieniuk and Le Corre, 2013).

In Europe, the directive 2009/128/EC on the "Sustainable Use of Pesticides" was implemented to promote Integrated Pest Management (IPM) in conventional farming systems. The directive's aim is to limit the use of pesticides to reduce their impact on human health and the environment by establishing alternative non-chemical control methods (Moss, 2010). Based on this approach, weed control should be performed according to the concept of Integrated Weed Management (IWM). IWM is a holistic plant protection approach. With regards to weed control, it considers the complexity of weed communities by combining different management techniques with a focus on preventive weed control measures. This means that unfavorable growing conditions must be created for weeds throughout the farming year (Buhler, 2002). Direct control measures like mechanical weeding can then be used as an independent procedure for weed removal. Mechanical weed control can also be combined with variable rate herbicide application or by restricting herbicide application to band spraying so that only the crop rows are treated (Harker and O'Donovan, 2013). Implications of the IWM measures are, however, an increased system complexity which might impede their adoption by farmers (Bastiaans, Paolini and Baumann, 2008).

The efficiency of mechanical and chemical weed control systems varies with the weather, soil, and crop conditions and the skill of the farmer (Bowman and Outreach, 2002). However, especially mechanical weeding requires a lot of experience and a careful execution of the treatments to prevent crop damage. Especially implement steering mistakes due to careless driving or due to driver fatigue can substantially decrease yields (Home *et al.*, 2002). Therefore, manually steered weeding tools always have the disadvantage that the efficiency and precision of the treatment is limited to the person guiding the implement. Furthermore, it is an exhausting task and often requires two workers, one for driving the tractor and the other for guiding the tools along the crop rows. Facilitating this work by implementing sensor-based guidance systems can be an option to make mechanical weeding more efficient and economically attractive to farmers.

The implementation of sensors in agriculture is part of the precision agriculture (PA) concept. PA focuses on the gathering and processing of information to create a sustainable agricultural production with increased resource use efficiency and quality (*International Society of Precision Agriculture*, 2019). Typical sensors that are used in agricultural environments for PA applications are satellite, aerial, handheld or tractor-mounted systems (Mulla, 2013). Especially direct weed control measures for mechanical weeding profit from the combination of sophisticated machine guidance with new weeding tools and techniques (Bond and Grundy, 2001). For example, camera-guided hoeing has already been established in wide row crops such as sugar beet, maize and soybean during the past years. The performance of camera-hoeing has been extensively tested in several field experiments (Tillett *et al.*, 2003; Wiltshire, Tillett and Hague, 2003; Kunz, Weber and Gerhards, 2015; Kunz *et al.*, 2016, 2018). Even single crop plant treatments can be made possible with camera-based steering systems (Slaughter *et al.*, 1999; Gobor, 2013). Camera-guided implements combine the advantages of increasing the weeding precision and the driving speed compared to manual steering. This saves the farmer time and facilitates the work. Other sensors including the global positioning system (GPS), laser and ultrasonic sensors are also frequently used for agricultural applications. Agricultural robots have also been developed in the past years. They usually have a combination of camera-vision, laser and ultrasonic sensors. However, robots also rely on camera-vision for crop plant or crop row detection whereas laser and ultrasonic sensors are mainly used for general navigation across the field to detect row structures and to avoid collision with other obstacles. Currently, only a few robots are commercially available. Further developments must be undertaken to increase the performance in an outdoor environment for effective and reliable autonomous mechanical weed control measures.

1.1 Objectives of the Thesis

The objectives of this thesis were to i) collect and analyze the information of current sensor-based mechanical weed control applications and to present an outlook on possible developments, ii) investigate the impact of interrow hoeing in narrowly spaced cereals on the weed control efficacy and crop yield and iii) develop and test a new weeding tool for intrarow mechanical weed control in sugar beet with autonomous weeding systems.

1.2 Structure of the Thesis

This thesis comprises **three scientific papers** which were published in the peer-reviewed journal *Agriculture* and *Computers and Electronics in Agriculture* or submitted to the peer-reviewed journal *Weed Research.*

The first paper is a review titled "**Sensor-based mechanical weed control: Present state and prospects**" and was published in the journal *Computers and Electronics in Agriculture.* An overview of the developments in sensor-based mechanical weed control is given. Furthermore, it provides a context for the practical applications of sensor-based mechanical weeding, possible issues, and an outlook for future developments.

The second paper is titled "**Adjustment of weed hoeing to narrowly spaced cereals**" and was published in the journal *Agriculture.* It describes two field trials concerned with interrow hoeing in narrow cereal row spaces. The effects of hoeing on the weed control efficacy and the cereal yield were analyzed.

The third scientific article is titled "**Sensor-based intrarow mechanical weed control in sugar beets with motorized finger weeders**" and was submitted to the journal *Weed Research.* It describes the development and testing of a new intrarow weeding tool designed for the use with robots performing mechanical weeding in wide row crops like sugar beets.

The findings of each paper are discussed in the section General Discussion. A comprehensive Summary of the entire thesis is provided after the General Discussion.

2 PUBLICATIONS

2.1 Sensor-based mechanical weed control: Present state and prospects

Jannis Machleb[a,*], Gerassimos G. Peteinatos[a], Benjamin L. Kollenda[a], Dionisio Andújar[b], Roland Gerhards[a]

[a]*Weed Science, Institute of Phytomedicine, Otto-Sander-Str. 5, Stuttgart, 70599, Germany*
[b]*Centre for Automation and Robotics, CSIC-UPM, Arganda del Rey, Madrid 28500, Spain*

Published in: *Computers and Electronics in Agriculture, 176, 105638.*

The original publication is available at:
https://doi.org/10.1016/j.compag.2020.105638

2.1.1 Abstract

Mechanical weed control in agriculture has advanced in terms of precision and working rate over the past years. The real-time communication of implements with sensor systems further increased the potential of mechanical weeding. There is a wide array of available sensors including image analysis by camera, GNSS, laser and ultrasonic systems that can improve weed control efficacy in combination with mechanical systems. Every sensor type has its advantages and disadvantages. Camera-steered hoes with a hydraulic side shifting control for row crops are robust and reliable and they are now widely available from different manufacturers. The robotic sector has difficulties handling the everchanging environment of agricultural fields and fully automated systems may not exist for many years. This review looks at the developments in sensor-based mechanical weeding since the 1980s. It focused on scientific studies presenting data results of their work, with the aim of providing an overview of the possibilities of sensor-based systems and to show their efficacy. The practical application for current and future farms is considered and discussed. Furthermore, an outlook towards sensor-based mechanical weeding in the future and necessary improvements are given.

Keywords: mechanical weed control, sensors, plant recognition, machine vision, guidance

*Corresponding author.

Email addresses: jmachleb@uni-hohenheim.de (Jannis Machleb), G.Peteinatos@uni-hohenheim.de (Gerassimos G. Peteinatos), benjamin.kollenda@uni-hohenheim.de (Benjamin L. Kollenda), d.andujar@csic.es (Dionisio Andújar), Roland.Gerhards@uni-hohenheim.de (Roland Gerhards)

2.1.2 Introduction

Mechanical weed control has been a sustainable option for weed removal throughout farming history. However, controlling weeds with mechanical means is challenging and requires the combination of different weeding techniques and cultivation strategies to achieve economically acceptable weed control levels (Oriade and Forcella, 1999). Additional challenges include the high weather dependency and the slow working speed of mechanical weed control operations compared to a conventional sprayer. The scientific literature reports on various positive and negative results of studies dealing with mechanical weed control. Wiltshire, Tillett and Hague (2003) and Kunz, Weber and Gerhards (2015) found that mechanical weeding alone or in combination with herbicides produced similar sugar beet yields as conventional spraying. Rasmussen (2004) also achieved positive yield results in cereals for mechanical weed control whereas (Lötjönen and Mikkola, 2000) observed that mechanical weeding could also reduce grain yield or show no effect at all.

Despite its complexity, mechanical weeding should be viewed as a valuable complement to chemical weed control. Today, the main drivers for the introduction of mechanical weed control include the global spread of herbicide-resistant weeds and also the lack of herbicides for various crops (Busi *et al.*, 2013). Herbicides became a limited resource because only a few herbicides with a new mode of action were discovered during the past 20 years (Duke, 2012). With currently 262 (152 dicotyledon and 110 monocotyledon) herbicide-resistant weed species globally (Heap, 2020), alternative weed control methods and effective resistance management practices, as required for Integrated Pest Management (IPM), are essential to further decrease herbicide usage (Harker and O'Donovan, 2013). Furthermore, political trends must be kept in mind. The new EU restrictive legislation laid the groundwork for the aim of reducing agricultural pesticide use by 50% in Europe by the year 2030. Until then it is important to find suitable alternatives that can compensate the loss of synthetic herbicides. Finding suitable alternatives for fungicides and insecticides is also of great importance, however, this is not the focus of this work.

In contrast to spraying, mechanical weeding involves the repetitive, direct physical contact of tools with the soil. There is no straight answer to the question of how many passes are necessary with a mechanical implement to maximize weed control whilst minimizing crop damage. The treatment amount (frequency) and intensity depend on several factors including the crop growth stage, weed growth stage, weed density, row spacing and the soil conditions (Melander, Rasmussen and Bàrberi, 2005; Van Der Weide *et al.*, 2008; Kolb and Gallandt, 2012). However, some general assumptions can be made about different crops. For example, sugar beets, soybean and maize have a wide row spacing and usually require between 2 and 3 mechanical treatments (Gunsolus, 1990; Kunz *et al.*, 2018). In competitive crops like cereals, where the row spacings are usually small, 1–2 passes with the harrow can suffice to achieve a sufficient weed control efficacy (Brandsæter *et al.*, 2012). Mechanical weeding always poses the risk of partial crop damage or even complete crop plant loss. Physical damage to crop plants may stunt plant growth and allows pathogens to enter which may lead to secondary infections and also decreases the yield (Findlay *et al.*, 1996). For example, sugar beet roots can be infected with Fusarium- or Phoma root rot if they suffer mechanical damage (Draycott, 2008, p. 304). Due to the risk of imminent crop damage, exact guidance of the mechanical tools is essential.

But despite all modern sensor technology, the basic concepts of mechanical weeding must be kept in mind to identify the parameters that can be improved for a higher weed control efficacy and ease of use for the operator. Factors that need to be considered for effective mechanical weed control are the treatment timing, frequency, cultivator type and intensity (aggressiveness). The crop growth stage, soil texture and moisture as well as the preceding and subsequent weather conditions must also be taken into account. Especially the timing of treatment is an often-discussed topic among farmers but general assumptions about the timing and frequency of a mechanical weeding operation are difficult to make since this is highly location specific (Oriade and Forcella, 1999). Farmers generally agree that the earlier mechanical treatments are applied, the greater is the weed control effect and the resulting yield. Several studies support the importance of early weed control in different cropping scenarios including dry beans (Vangessel *et al.*, 1998), onions (Bond *et al.*, 1998) and winter wheat (Welsh *et al.*, 1999; Knezevic *et al.*, 2002; Melander *et al.*, 2003). Knezevic *et al.* (2002) also highlight this issue by providing a general concept on the critical period for weed control. However, a study by Rasmussen *et al.* (2010) conducted in spring barley indicated that different treatment timings within a 2 week window did not affect crop yield and it was concluded that adjusting the harrowing intensity to the crop growth stage was more important than the timing of post-emergence harrowing. All studies share the view that the type of cultivation measure has a great influence on the weed control efficacy.

The three cultivation categories for post-emergent mechanical weed control in agricultural row crops are:

1) **whole field treatments,**
2) **inter-row, and**
3) **intrarow treatments.**

Whole field treatments refer to methods which treat the entire field, whereas inter-row and intrarow mechanical weeding leaves the close- to-crop area untreated. Harrowing (Figure 2.1.2-1) or rotary hoeing are typical implements for whole field treatments, where damage to a certain percentage of crop plants is expected and tolerated. Harrowing is suitable for various crops including cereals, maize, peas, and soybean. Some crops (e. g. sugar beets) are less suited for harrowing due to their delicate leaves and may suffer yield losses (Ascard and Bellinder, 1996). In general, harrowing in sugar beets is possible but it requires an in- creased seed density and it should not be performed before the true 4 to 6-leaf stage of the beets (De Buck *et al.*, 1999; Van Der Weide *et al.*, 2008; Cioni and Maines, 2010). Therefore, it can be used in addition to hoeing but since sugar beets need intensive weeding even prior to the 4-leaf stage, a reliance on harrowing alone could be inadequate for effective weed control. Inter-row treatments (e.g. hoeing) cultivate the space between the crop rows (Figure 2.1.2-1) and intrarow weeding refers to treatments of the crop row. Usually a combination of inter-row and intrarow treatments is applied simultaneously to maximize the treated field area. A typical set up for maize or soybean would be a hoe with goosefoot

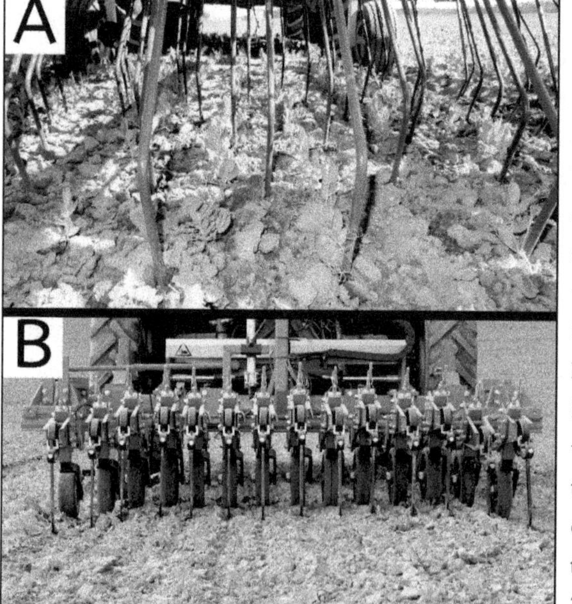

Figure 2.1.2-1: Harrowing peas (A) and hoeing in soybean (B).

sweeps in combination with finger weeders. Finger and torsion weeders are specialized tools which are very effective at removing weeds from the crop row (Riemens *et al.*, 2007; Van Der Weide *et al.*, 2008).

Keeping this in mind, it is clear that sensor-based (intelligent) cultivators must be able to either identify crop row structures or single crop plants, to ensure an optimal alignment of the tools with the crop. The mechanical tools should be guided as close to the crop plants as possible to increase the treated field area. However, physical crop damage must be avoided. Crop plants are easily damaged due to steering or guidance errors, which can lead to considerable yield losses (Home *et al.*, 2002; Melander, Rasmussen and Bàrberi, 2006). This stresses the importance of accurate cultivator guidance. The closer the weeding implements are guided along the plant rows, the more important precise steering becomes (Melander and Hartvig, 1997).

Research information and the resulting conclusions must be transferred into production systems. Therefore, Table 2.1.2-1 summarizes 9 basic requirements that must be fulfilled for half or fully automated weeding systems to be robust and efficient in typical farming environments. After all, higher digitization also corresponds with higher acquisition costs and utilizing modern technology must combine certain new advantages to be ecologically and economically profitable.

Table 2.1.2-1: Basic requirements for efficient sensor-guided mechanical weeding machines.

- Avoidance of crop plant damage by increasing the precision of the implement guidance and correctly identifying crop rows or single crop plants in different growth stages
- Cope with varying field characteristics such as terrain slopes, different soil types, high weed pressure, crop row gaps (missing crop plants), and varying crop appearance (e.g. crop height and red vs. green plants)
- Economic advantages must be present to compensate for higher acquisition costs
- Enable less experienced operators to perform advanced weed control tasks (e.g. hoeing close to a crop row)
- Ensure a consistent working depth of the tools in the soil
- Extend operation time from daylight to nighttime
- Facilitate the workload by automatically guiding the implement(s)
- Increase ground coverage by enabling higher driving speeds
- Robustness of all system components to withstand varying field, weather and illumination conditions including dust, water, temperature variations, machine vibrations, lateral and horizontal forces that may arise during the application of treatments

This study focuses on smart technologies where sensors are used for the detection of weed or crop structures to perform mechanical weeding. Where possible, experiments that provide solid facts on the performance of specific methods and technologies are included. The discussion pinpoints challenges of current and future mechanical weed control technologies. This review has been divided into four sections to gain a better overview of the relevant possibilities for sensor-guided mechanical weed control. The sections are (I) Autonomous Robots and (II) Half Autonomous Tractor-pulled Implements which are divided into (IIa) Machine Vision, (IIb) GNSS Sensor Systems and (IIc) Laser and Ultrasonic Sensors.

2.1.3 Autonomous Robots

Agricultural robots can be used for many different field applications including harvesting or weed control. Their appearance and set up may vary and can range from a modified tractor to a small, specialized platform which travels the field autonomously to execute a crop operation (Emmi *et al.*, 2014). Robotic systems have been an essential part of industrial style work lines for many years (Day, 2011). But their implementation in food and farming systems has been troublesome because of the complex, unstructured and everchanging environments as illustrated in Table 2.1.3-1 (Nof, 2009; Day, 2011; Hiremath *et al.*, 2013; Bechar and Vigneault, 2017; van Henten *et al.*, 2017).

Table 2.1.3-1: Robots in agriculture and other environments, adapted and modified from Bechar and Vigneault (2017).

	Environment	
	Structured	Unstructured
Structured objects	Industrial domain	Military, space, underwater, mining domains
Unstructured objects	Medical domain	Agricultural domain

Even though agricultural robots do not have to be as accurate as industrial robots on a cm vs. mm scale, they must cover large areas and are often faced with deviating terrain slopes and irregular field sizes (Edan, 1995). Furthermore, the exact location of and the distinction between weed and crop plants is a crucial prerequisite for robotic systems to perform mechanical weeding (Baerveldt and Åstrand, 2002). This task is hindered by variations in plant appearance, alternating weed pressures and weed populations from field to field during the entire cropping season and must be tackled with highly developed software solutions (Midtiby *et al.*, 2016). The complexity of creating robots for agricultural applications was well addressed by Bechar and Vigneault (2017): in contrast to industrial uses, agriculture is the only domain where unstructured objects coincide with an unstructured environment. Therefore, creating autonomous mobile robotic systems for agricultural applications first became a realistic aspect of scientific studies with the advancements made in sensor-supported field machinery in recent years (Ruckelshausen *et al.*, 2006).

Concerning weed removal with autonomous mobile robots, Slaughter *et al.* (2008) specified the following technical requirements as the essential characteristics: self-guidance, weed detection, and identification, precision intrarow weed control, and weed mapping. Of those characteristics, plant detection, identification and determination of the exact position are the most challenging tasks. To locate crop or weed plants, a series of steps are necessary and there are two main concepts of using robotic systems for plant care and mechanical weeding. The first one is geo-referencing the planted or seeded crop plants via GNSS (Global Navigation Satellite System) and storing their exact location in a plant map (Pérez-Ruiz and Upadhyaya, 2012). A robotic system can later use this information to locate each crop plant and perform the necessary weeding tasks. However, the disadvantages are that errors during seeding or planting are going to affect the performance of the weeding process. If weeding is solely based on GNSS information, irregularly spaced crop plants are not considered and will be damaged or killed. Planting can be an option to avoid dislocation of the seeds during seeding, but it is not always possible or economically feasible. Studies concerned with the creation of seed maps showed that the location of mapped crop seeds can be within 34 mm (Ehsani *et al.*, 2004) or 16–43 mm (Griepentrog *et al.*, 2005a) of the germinated crop plants. These results indicate that creating seed maps can be feasible for accurate weeding operations. Still, a more advanced concept is seen in the combination of machine vision and GNSS-navigation. The robotic system receives the necessary GNSS coordinates for general field navigation, but it must be able to detect crop plants or rows on its own without a-priori-knowledge in terms of a seed map. Camera-guided implements can then perform mechanical weeding on the exact location of the crop plants or along crop rows.

Studies of using GNSS-referenced crop plants for mechanical weed control can be found in Nørremark *et al.* (2008, 2012). Their robotic system was able to perform mechanical inter- and intrarow weed control while traveling along crop rows with a cycloid hoe. It was found that in combination with the intrarow treatment of the cycloid hoe, up to 91% of the field area could be treated. This represents a large amount of field area which in turn could result in a high weed control efficacy. Similarly, Bakker *et al.* (2010c), Bakker *et al.* (2010a), Bakker *et al.* (2010b) described the application of an Intelligent Autonomous Weeder (IAW, Wageningen University) for inter-row hoeing in maize with RTK- GPS (Real Time Kinematics Global Positioning System) referenced seeding. The authors state that no crop plants were damaged at a driving speed of 0.5 m s^{-1}. Both studies demonstrated that autonomous mechanical weeding is possible with GNSS-derived data. However, the mean working speed of 0.5 m s^{-1} is not as fast as a conventional tractor would work. The slow working speed could be compensated by the robotic system working day and night. However, another problem arises: usually, much higher driving speeds of at least 4 km h^{-1} are needed for effective mechanical weed control with conventional tractor-pulled implements for mechanical weeding. Some implements such as harrows even require driving speeds of at least 6 to 12 km h^{-1} to provide the best weed control efficacy (Bowman and Outreach, 2002). The reason is that large proportions of weed plants are not only cut or uprooted by hoeing devices, but also a certain amount of small weed plants are buried with soil, and their growth is impeded. This implies that effective weeding implements for robotic systems should be of a rotating or similar actively engaging nature because robotic systems cannot generate high enough speeds for the soil burial effect to happen due to their comparatively slow working speed. Especially rotating tools are very powerful, and they even kill large weed plants which hoes, or harrows may not be able to eliminate. Thus, rotating implements are the ideal solution to compensate for the speed limitations that robotic systems have.

The BoniRob platform by Bosch Deepfield Robotics has such an active tool to perform mechanical weed control (Langsenkamp *et al.*, 2014). A stamp is positioned above a weed plant and is then lowered onto the plant to destroy it. This might work in light soils, but difficult soil conditions and the slow working concept will be problematic. The tube-stamp is therefore not seen as a promising method for mechanical intrarow and close-to-crop weed control despite its weed control efficacy of up to 94%. Nørremark *et al.* (2012, 2008) addressed the issue of finding a suitable tool for mechanical weeding with robots by developing and testing a cycloid hoe for intrarow treatments. However, limiting their navigation ability to GNSS derived data and excluding additional features, such as machine vision, does not entirely satisfy the requirements of an autonomous field unit.

Instead of using GNSS-reference data, Baerveldt and Åstrand (2002) equipped a robotic system with a forward-looking camera working with a grey level vision system and a near-infrared filter. The system was able to locate sugar beet rows and identify single sugar beet plants. After identifying the position of each sugar beet plant, a mechanical tool consisting of a rotating wheel was lowered by a pneumatic cylinder. The wheel rotated perpendicular to the crop row, thus eliminating weeds around each sugar beet. This setup demonstrated how robotic systems can cater to single crop plants. Further improvements should provide the robot with the ability to access a nearby charging station on its own if power levels reach a critically low level.

Apart from weed control in a field, robots may also be used in pastures to eliminate certain weed species, thereby decreasing the required manual labor (van Evert *et al.*, 2011). An autonomous robot was developed and tested by van Evert *et al.* (2011) to find and eradicate broad-leaved dock (*Rumex obtusifolius* L.) on a commercial farm. Navigation was achieved with a predefined path on which the robot navigated via GNSS. A downward-looking camera then identified *R. obtusifolius* plants and a mechanical tool was positioned over the center of the weed plant. The tool was lowered onto the plant and rotating knives cut the plant to pieces. Van Evert *et al.* (2011) reported a successful weed detection rate of 93% and the successful removal of 75% of those plants. Large weeds, such as *Cirsium arvense* L. or *Rumex* spp. L. are often a problem in pastures. Manual labor is required to remove those plants by hand or by spraying them with an herbicide. Being able to rely on automatic mechanical weeding systems for removal of such weed species would be of great environmental importance and it would drastically decrease manual working hours. By using more than one autonomous system per field the working time could be decreased further. Thereby the treated pasture area would increase immensely.

Based on this idea, Noguchi *et al.* (2004) devised the concept of a fleet of robots according to a master–slave system. In such a system several autonomous vehicles execute all agricultural work related to the crop from sowing until harvest. A master-robot incorporates the function of decision making and planning, while one or more slave-robots follows the master-robot and assist in the assigned tasks. The RHEA Project (Robotics and associated High-technologies and Equipment for Agriculture) took small market available tractors and modified them to perform agricultural tasks on their own. Supervision is provided by a multi-level architecture with several supervision levels. The tractors were equipped with machine vision for weed and crop row detection, laser range finders for obstacle detection and GNSS for general navigation in the test field (Emmi *et al.*, 2014). Pérez-Ruiz *et al.* (2015) equipped one of the RHEA project tractors with a mechanical-thermal weeding tool. While the thermal tool was machine vision assisted and selectively eliminated intrarow weeds in maize, the mechanical tools performed continuous inter-row hoeing regardless of weed cover. A detailed description and images of the mechanical implements can be found in Gonzalez-de-Santos *et al.* (2017). The mechanical-thermal combination achieved weed reduction results of 90% and no major crop losses were recorded. Besides maize, the RHEA-fleet-system has been tested in other crops such as onion and garlic (Conesa-Muñoz *et al.*, 2015). At Harper Adams University the “hands-free hectare project” (Hands free hectare, 2018) followed the same concept. In 2017 it was the first project to fully automate a single crop cycle in spring barley from sowing until harvest. No mechanical weed control was performed, but more results can be expected in the future.

The AgBotII (Bawden *et al.*, 2017) has been developed as another agricultural robot combining mechanical and chemical weed control implements. Individual weeds were classified and either removed mechanically or treated with a chemical compound. Classification accuracies of 90% were achieved. According to the authors, weed removal with the mechanical tools was successful in removing wild oats and sowthistle. By combining mechanical and chemical weed control strategies robots may become a powerful alternative to conventional control techniques.

Eventually, such systems may be used in everyday farming situations. However, until then major work is required. The biggest problem is to guarantee a continuous operating state of the robot. Therefore, robotic manufacturers should focus on providing adequate customer support in case of a malfunction. The support should consist of online solutions as well as local technicians who could quickly act and perform necessary on-farm repairs. Today even tractors are equipped with high-end technology and often the farmer is no longer able to perform necessary repairs. During critical time periods (e.g. weeding, harvesting) the farmers need the maximum machine uptime and availability of the relevant equipment. Possible failures need to be fixed as soon as possible, which can be hindered by the lack of available personnel and know-how. The maintenance of robotic units will not be any less easy. However, despite the obstacles robots have to overcome in an agricultural environment, their potential is being increasingly developed and can help to reduce environmental costs in conventional precision farming, and the required human labor in organic farming (Gobor, 2013; Chen *et al.*, 2016). Since robots are usually small vehicles, they can easily be powered with solar panels and electricity which eliminates the dependency on fossil fuels. Furthermore, robots can work 24 h a day and are able to replace human labor even if they are slower working than a skilled worker.

Several companies are working on autonomous solutions for mechanical weeding to keep the input of herbicides on the environment as low as possible. NAÏO (France) provides three models of robots ("OZ", "DINO" and "TED") for agricultural tasks, which can partially work autonomously (Autonomous weeding, agricultural robots - Naïo Technologies, 2019). However, navigation of their robots is mostly based on RTK-GPS navigation with a pre-planned path and still requires an elaborate setup. Similar to NAÏOs "DINO" robot model, the company Carré (France) is working on a related concept with the development of the "ANATIS" robot. It is also supposed to aid farmers in vegetable farming by performing mechanical weeding (Carré - Made for Agriculture, 2020). Also, small systems like the "Tertill" are developed for small scale weeding in gardens (Franklin Robotics | Home of Tertill, the robotic garden weeder that is powered by the sun!, 2020). Even though a small robot can only tend to a limited area they could also be used in a swarm of many robots to treat entire fields. An advantage would be that single units can be replaced quickly. Other robotic systems, which are still under development and not yet commercially available exist. This includes the PUMAgri by SITIA which is also being developed with the aim to reduce the herbicide input in agriculture (Platform PUMAgri - Sitia - Bancs d'Essais et Innovation Robotique, 2020).

From the previously mentioned robotic systems the sizes can vary between very small units (TerTill or OZ) to robots the size of a small tractor as in the RHEA Project. Much larger robots are not seen as an advantage because soil compaction would increase dramatically. The idea behind robot systems is that they should perform repetitive tasks 24/7 to maintain a low weed pressure. Therefore, agricultural robots must not be too heavy. Otherwise the negative effects due to soil compaction would outweigh the benefits of the mechanical weed control measures. Apart from the size and weight of an intelligent implement for weeding, sloping terrain is also an issue as mentioned in Table 2.1.2-1. The company Energreen explicitly created the RoboZERO robot for weeding and mowing which can cope with slopes of up to 30° (Robozero Weeding Robot, 2020). However, it is controlled via a remote control by an operator and does not fully work autonomously. A system which combines all the prerequisites for successful sensor guided mechanical weeding is the robot produced by FarmWise. Based on deep learning, their robot captures and analyzes plant images to detect and remove weeds (FarmWise, 2020). It seems to be a very robust system that can be used for long-term field work in many conditions. A similar but solar powered autonomous robot is the "Farmdroid FD20" which can perform sowing and hoeing between crop rows. By combining sowing and weed control, an interesting hybrid robot has been created which can contribute to the general concept of autonomous farming. According to their website, the Farmdroid can care for 20 ha per season

(FarmDroid FD20, 2020). It is a slow working robot (1 km h^{-1}) but this can be compensated because it works continuously.

2.1.4 Half Autonomous Tractor-pulled Implements

2.1.4.1 IIa Machine Vision Systems

The benefits of machine vision for industrial uses also led to in- creased research in a variety of agricultural applications such as crop row following (Åstrand and Baerveldt, 2005) and automatic guidance of robotic systems (Baerveldt and Åstrand, 2002; Pérez-Ruiz *et al.*, 2015; Gonzalez-de-Santos *et al.*, 2017). However, the following factors can negatively impact image acquisition and processing in the field: (1) Changing environmental conditions, (2) varying plant sizes and shapes, (3) missing crop plants in a row (Åstrand and Baerveldt, 2005), (4) contamination of plants by foreign particles and thereby changing their appearance (Montalvo *et al.*, 2013) and (5) diverging amounts of weeds present in the field (Slaughter *et al.*, 2008).

Unfavorable conditions for image acquisition and the ideal placement of the camera in relation to the mechanical weeding implement were issues that had to be explored in the very early studies of agricultural machine vision. Several authors including Guyer *et al.* (1986) and Reid and Searcy (1988) described implications due to environmental factors or technical issues (e.g. lack of robustness) in their studies. Guyer *et al.* (1986) also concluded that for mechanical weed control systems, it would suffice to differentiate between crop and no crop (weeds) only. This was an important statement since it already suggested to keep the technology as simple as possible and that the focus of machine guidance should rest on recognizing the most prominent structure inside a field: the crop rows. By combining machine vision with radio navigation, the precision of the weeding operation can be increased further (Tillett, 1991). It was also observed that major issues arose from mounting a camera system in the front of the tractor because the distance to the cultivator in the rear was too far and complicated the implement guidance (Billingsley and Schoenfisch, 1995, 1997). The risk for alignment errors increases with the distance between the camera and the cultivator.

Another issue for machine vision in an open field is lighting and the visual appearance of the crop or weed plants. This can be a variation in color between different crops or between the same crop type (e.g. red and green cabbage). The software algorithm should also be able to distinguish between different shades of the same color (e.g. green) to fine-tune the crop recognition result (Gerrish *et al.*, 1997). Shadow casting behind the tractor in the view field of the camera during intense sunshine is another problem. This can be avoided by placing a cover over the camera of the cultivator to prevent the negative influence of direct sunlight during image acquisition (Slaughter *et al.*, 1999). Adding an artificial light source ensures constant illumination with the same intensity.

The first vision guidance systems focused on crop row recognition for inter-row hoeing. The easiest way of changing the alignment of a hoe with the crop rows is the use of a hydraulic cylinder for lateral adjustment to push the hoeing frame to the left or right (Slaughter *et al.*, 1999). This represents a fast and accurate application and is easily established in wide spaced crops like sugar beets or maize. However, difficulties arise when the row spacing decreases. Tillett *et al.* (1999) presented a machine vision system which was able to work in 22 cm row spacing. The authors stated that the accuracy could be increased by installing a second camera to monitor additional crop rows. In a follow-up experiment with sugar beet rows, Tillett *et al.* (2002) tested the system under unfavorable conditions with large gaps inside the crop rows. It was observed that row gaps of 4 m were not an issue for the vision system and steering continued without flaws. Additional experiments in wheat can be found in Home *et al.* (2002). Their results showed that the vision system improved the accuracy and speed of mechanical weed control in cereals. The developed system has been brought to the public markets and it is commercially available as the Garford Robocrop steering system. Further insights into the setup and specifications of the Garford Robocrop system are presented in Connolly (2003).

Wiltshire *et al.* (2003) performed field trials with the Garford Robocrop in sugar beets and tested the row guidance accuracy. The side shifting frame could move ± 15 cm left or right to stay allocated with the crop rows. The overall working width was 6 m and 12 rows of sugar beets were hoed at once. The driving speed for all treatments was 5 km h^{-1}. The results were promising and showed band spraying plus vision-hoeing to be equal or better than the overall herbicide application. The sugar beets were not damaged during the experiment and the yield was not decreased by the mechanical treatments. Wiltshire *et al.* (2003) also addressed the importance of novel methods in tool shapes concerning sensor-guided mechanical weed control. The conventional tool designs might be unfit for higher speeds due to excessive soil throw onto the crop plants.

Kunz *et al.* (2015) conducted field experiments in sugar beets and soybean. Machine vision steering was compared to RTK-GPS assisted steering, manual steering, and conventional herbicide applications. Goosefoot blades were used for inter-row hoeing and finger weeders treated the intrarow area. The automatic hoe guidance led to 89% and 87%

of weed plant reduction in soybean and sugar beets respectively. With automatic steering, driving speeds could be increased from 4 km h^{-1} to 7 km h^{-1} and 10 km h^{-1} without negatively affecting the crop. Continuative field experiments in sugar beets and soybean recorded similar results of 82% weed reduction when mechanical weed control was combined with a camera-steered hoeing frame (Kunz *et al.*, 2016). Experiments in maize with a camera-steered hoe led to average weed density reductions of 85% compared to the untreated control (Kunz *et al.*, 2018).

Tracking of crop rows for inter-row weeding had been successfully established in the early 2000s and the next step was to address the problem of intrarow weeds. Therefore, Tillett *et al.* (2008) equipped the Garford hoeing system with newly designed cultivation blades. The blades were discs with a section cut out and resembled a half-moon shape. Guided by a vision system for single crop plant identification the discs rotated around a vertical axis and performed intrarow weed control in cabbage. The cut-out section allowed for cabbage plants to remain unharmed. Results showed an intrarow weed reduction of up to 87%. Fennimore *et al.* (2014) tested the hydraulically controlled rotating discs of the Garford InRow Weeder in bok choy, celery, lettuce, and radicchio. With the rotating cultivator, 25% less time was spent for additional hand weeding and thinning of the seeded lettuce. Weed densities in transplanted crops were reduced by 85%. Now the Garford "Robocrop Guided Hoes" and the "Robocrop InRow Weeder" are commercially available for a variety of agricultural row crops.

Another trial with a commercial system for intrarow weeding has been performed by Melander *et al.* (2015) with the Robovator (F. Poulsen Engineering, Denmark) in transplanted onions and cabbage. The Robovator consists of a machine vision system for single crop plant identification and one pair of tines with knife-like blades for each crop row. The tines open when a crop plant is encountered and close after passing the crop plant. Thus, weeds are removed mechanically from the intrarow space. A safety zone around the crop plant can be defined based on the user's needs and the crop plant's growth stage. Melander *et al.* (2015) found no distinct weed control effects between the smart cultivator (Robovator) and conventional treatments (torsion weeder and harrowing). Continuing studies with the Robovator were conducted by Lati *et al.* (2016) in direct-seeded Broccoli and transplanted lettuce. At moderate to high weed densities the Robovator decreased hand-weeding hours by up to 45%. Between 18% and 41% more weeds were removed by using the Robovator than with standard cultivator knives. These studies demonstrated that smart intrarow weeding can be feasible, and it is definitely of great interest for organic farmers to reduce manual weeding hours.

2.1.4.2 IIb GNSS Sensor Systems

The global navigation satellite systems (GNSS) is a navigation system that uses signals provided from space satellites which transmit positioning and timing data to GNSS receivers (European Global Navigation Satellite Systems Agency, 2014). Currently, the four global systems GPS (USA), GLONASS (Russia), Galileo (EU) and BeiDou (China) exist. It is found in a variety of applications in many fields of research such as in precision agriculture to provide geo-referenced soil and yield data as well as allow for variable rate application of herbicides (Auernhammer and Muhr, 1991; Tyler *et al.*, 1993; Borgelt *et al.*, 1996). However, as pointed out by Hiremath *et al.* (2013), GNSS by itself can have insufficient accuracy for certain tasks and when the signal is interrupted navigation can fail. To overcome the accuracy problem, precision agriculture requires the utilization of more accurate methods, typically RTK-GPS technology, which uses a local reference station (e.g. on the farm) to correct and refine the signal received from the satellites, thus offering sub-centimeter level accuracy (Keicher and Seufert, 2000; Zhang, Wang and Wang, 2002).

Rasmussen *et al.* (2012) investigated the possibility of mechanical intrarow weed control in geo-referenced sugar beets. The tines of a cycloid hoe were guided with RTK-GPS in relation to the previously recorded location of the sugar beets. The RTK-GPS guided cycloid hoe did not achieve significantly different results in terms of weeding selectivity when compared to mechanical weeding without GNSS-guidance. Furthermore, the overall crop damage was rather high. Pérez- Ruiz *et al.* (2012) performed mechanical weed control in geo-referenced tomato plants. In contrast to Rasmussen *et al.* (2012), two knife blades per intrarow space were used to eliminate weeds instead of a rotating tool. The blades opened when the position of a geo-referenced tomato plant was encountered and they closed in-between crop plants, thus cultivating the soil by leaving a safety zone around each crop plant. Thus, the weed control concept was similar to the Robovator (F. Poulsen Engineering, Denmark), except that no machine vision was used. None of the tomato plants were damaged, however, the study did not investigate important plant protection parameters such as the weed control efficacy because weed plants were not included in the assessment. Due to the high accuracy it can be a promising way of removing intrarow weeds.

New technologies in GNSS seeding also allow different tactics for mechanical weed control. Commercial attempts have been made for geo-referencing the position of individual crop seeds during seeding. For example, the GeoSeed by Kverneland (2020) aims at sowing crop plants with the same spacing from plant to plant. This would allow for hoeing to be performed perpendicular to the crop rows. Hoeing in different directions enables the disturbance of weed plants in every possible way. Sometimes weed plants are bent into one direction during hoeing if they are not effectively killed. Performing a second hoeing treatment perpendicular to the first treatment can eliminate weeds which might have survived the first pass with the hoe.

2.1.4.3 IIc Laser and Ultrasonic Sensor Systems

Ultrasonic- and LiDAR (light detection and ranging) sensors are also referred to as "distance sensors" and are usually combined with other sensors for navigation of vehicles (e.g. robots) in the field. Ultrasonic sensors can measure the distance from a sensor to a plant based on sound waves whereas LiDAR sensors measure the phase difference between an emitted laser beam and the reflected one. Additionally, LiDAR sensors are more precise than ultrasonic sensors due to higher measurement frequencies (Fernández-Quintanilla *et al.*, 2018). Especially ultrasonic sensors are inexpensive, readily available and work accurately (99%) within a range of 100 mm to 10 m (Tillett, 1991). However, when used on their own, studies and applications in mechanical weed control are scarce for both systems. Plant detection with LiDAR is based on the fact that chlorophyll reflects near-infrared-light (NIR), thus green colors of vegetation lead to high reflection values when using a LiDAR sensor. Their main application can be found in determining the necessary amount of fungicide or fertilizer application depending on the tree canopy volume in orchards (Colaço *et al.*, 2018). Other uses for laser-based sensor systems may include the soil depth control of agricultural implements such as in mechanical weeding to ensure a consistent working depth (van der Linden *et al.*, 2008). Reiser *et al.* (2018) have successfully used a laser scanner to steer a robot between maize rows in the early growth stages (first, second and third leaf collar). Other studies concerned with the detection and classification of weed plants inside a crop field have also been successfully performed with LiDAR systems (Andújar *et al.*, 2013). Similarly, ultrasonic sensors can be used to detect patches of high weed infestation in crop fields (Andújar *et al.*, 2012).

Cordill and Grift (2011) devised a laser-guided implement which detected maize stalk features. The authors assumed that the unique geometry of maize stalks would suffice for differentiation between maize, broad-leaf and grass weeds. Therefore, weeds were removed in a non-specific fashion by a mechanical tool, which treated the entire area surrounding the maize plants. Every time a maize stalk interrupted the laser signal the position was recorded and the information passed on to the weeding implement. However, no weed removal data was provided since the metal tines of the implement did not engage in the soil. Operating the laser-guided system in areas with broad-leaf and grass weeds led to 23.7% of crop plants being fatally damaged in each scenario. In comparison, only 8.8% of maize plants were fatally damaged when no weeds were present. It can be concluded that the system needs further improvement to minimize crop losses.

Ultrasonic sensors have been used by Rueda-Ayala *et al.* (2015) in a harrowing experiment in maize. By changing the angle of the harrow tines, different working intensities can be realized in the field. A steep angle leads to a high (aggressive) intensity whereas a gradual adjustment is less intense on crop plants and weeds. The aim of their study was to adjust the tine angles depending on the number of weeds present to spare the crop plants the mechanical stress where no weeds needed to be eliminated. With an ultrasonic sensor mounted in front of the tractor, online weed detection was realized, and real-time adjustment of the harrow tines was performed. It was assumed that areas with high plant densities have a high degree of weeds inside, thus requiring aggressive treatment with the harrow. Lower density values ultimately lead to a gentler treatment. The authors reported an average weed control efficacy of 51% ranging from 22% to 90%. The system was proposed as an online-harrow adjustment system, yet calibration of the system in the current crop stage was necessary.

2.1.5 Discussion and Outlook

Plant protection averts direct negative influences acting on a plant community. Therefore, it is a measure of success in maintaining a healthy plant community and achieving local, crop-specific goals. To achieve these goals, modern agricultural machinery has been developed into precise and reliable instruments. However, implement guidance is always limited to the skills of the farmer. Therefore, combining agricultural tools with sensor systems is one of the most important ways to further increase the production efficiency and to facilitate the labor. Table 2.1.5-1 lists prospects that are to be expected for sensor-based mechanical weed control in the future. Additionally, the challenges that must be overcome are also presented. The findings are discussed further within this section.

Table 2.1.5-1: Prospects and future challenges of sensor-based mechanical weed control.

Prospects

- A shift from the conventional "entire field" approach to an individual plant scale where the use of synthetic herbicides is reduced without compromising the weed control efficacy
- Standardized protocols for implement-sensor communication
- Further increase of aerial and ground sensor systems to gather large amounts of localized field data which can be used for diverse plant protection measures in more accurate prediction models to ensure a sustainable production management worldwide
- Intelligent cultivators and robots have access to historical data which contain information on past growing seasons. This information will be integrated in the Decision Support Systems for possible weed strategies in the current growing season. For example, known hotspots of weed infestation can be targeted timely and efficiently
- Smart implements for mechanical weed control decrease the time effort to adjust a cultivator manually to the crop conditions on a field (e.g. a camera-steered hoe that can adjust automatically to different crop row widths)
- Transfer of basic weeding tasks to multiple robots (fleet / swarm)
- Robots possess multiple implements on the same platform (mechanical, thermal, electrical, and chemical weeding tools) and they have an algorithm which chooses a non-spraying technique over a chemical treatment wherever possible and feasible – the algorithm also takes into account that not all weeds need to be removed, the aim is a diversified weed community on every field

Challenges

- A better understanding of selectivity of mechanical intra-row weed control, defining variable close-to-crop safety areas as crop growth progresses and determining thresholds for tolerable crop soil cover
- Implementing new algorithms of Artificial Intelligence (AI) in differentiating crops and weed species during intra-row weed control
- Increasing reaction time of physical intra-row weed control actuators
- Increase of robustness and battery life of agricultural robots
- Ease of use, and intuitive interface for the operator (farmer) must be provided
- An interaction with experts for identifying plant species with a low recognition rate. Rare and invasive weed species can therefore be identified in this way and handled, either by taking appropriate precautionary measures or for increasing the biodiversity
- Reevaluate the complete field concept and adjust our cultivation systems to better facilitate robotic mechanization in a modern, diverse, and sustainable agroecosystem (mechanization 4.0)

The efficiency of a weed control operation corresponds to a highly selective weed control level. Rasmussen *et al.* (2008) defined weeding selectivity as the amount of crop damage in relation to weed control efficacy. It is mostly dependent on the utilized weeding implement. However, the driving speed and diligent set-up of the mechanical tools are also very important parameters for efficient and successful mechanical weeding. Hence, effective mechanical weeding should be fast, reliable, precise and cover as much field area as possible. Cloutier *et al.* (2007) stressed the importance of guidance systems in mechanical weed control because of the benefits that come with every additional centimeter of cultivated space. Today, the mechanical removal of weeds from the inter-row area can be performed without large constraints, while removing weeds from the intrarow area is still a challenging task (Griepentrog *et al.*, 2004; Cloutier *et al.*, 2007; Tillett *et al.*, 2008; van der Linden *et al.*, 2008; Van Der Weide *et al.*, 2008; Müter, Damerow and Lammers, 2014). Unfortunately, the removal of intrarow weeds is especially important because their negative impact due to resource competition increases the closer they are growing to the crop (Heisel, Andreasen and Christensen, 2002; Zimdahl, 2004). It is, therefore, inevitable to combine implements with sensors to exploit the full potential of modern machinery.

The significance of precise mechanical weeding is shown in Figure 2.1.5-1, adopted and modified from (Pérez-Ruiz *et al.*, 2012). The illustration shows that treating only the inter-row space (zone A) allows for weeds to emerge, grow, and proliferate in the intrarow space (Zone B and C). In an ideal case, the entire field area, composed of the intra- and inter- row area, as well as the close-to-crop area, should be subject to mechanical weeding. However, depending on the crop type this is not always possible.

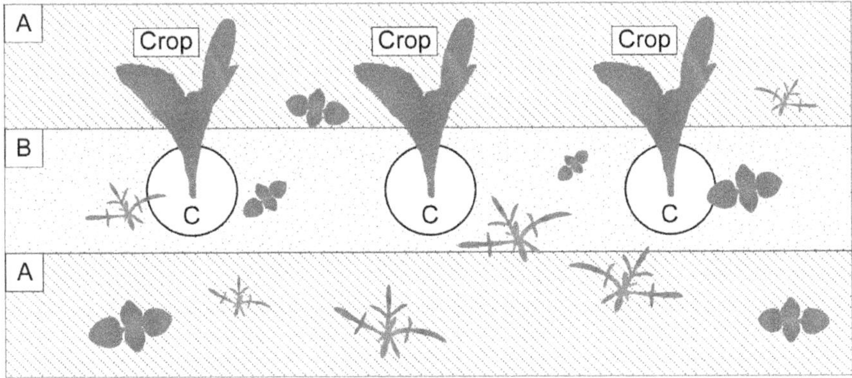

Figure 2.1.5-1: Depiction of the three weeding zones: A = inter-row space, B = intrarow space, C = close-to-crop space. Weeds are not labeled. Crop plants are labeled as “Crop”. Adopted and modified from (Pérez-Ruiz *et al.*, 2012).

Therefore, it is crucial to find the best sensor type combination for effective weeding while avoiding crop damage. Another important aspect is the continuous development of suitable end effectors. This can be an optimized size and shape for inter-row tools such as hoeing blades (Machleb *et al.*, 2018) or improved designs for intrarow weeding implements. In commercial mechanical weeding, SSWM (site specific weed management) is largely focused on camera-assisted steering of the implements and there is a great number of manufacturers who are equipping their hoeing machinery with cameras for crop row recognition (Fernández-Quintanilla *et al.*, 2018). Table 2.1.5-2 provides an overview of manufacturers for mechanical weed control systems that are available for purchase. Most of the implements are hoes which have the option to be combined with camera-steering. However, only a few manufacturers developed their own camera system and most of them rely on the CULTI CAM ('Claas CULTI CAM', 2020) or the Garford camera (Garford Farm Machinery Ltd | Precision Guided Hoes & InRow Weeders, 2020). Schmotzer (Maschinenfabrik Schmotzer GmbH, Bad Windsheim, Germany) has also developed its own camera system with the "Okio" camera (Reihenführung - Schmotzer Hacktechnik, 2020). So far, the Okio camera is only used by Schmotzer and no other companies.

Table 2.1.5-2: Overview of commercial sensor-based mechanical weed control systems in agriculture.

Sensor type(s)	Target zone	Manufacturer
Camera - Image analysis	Interrow cultivation	Agrokraft Carré Einböck Garford Farm Machinery Ltd. Gothia Redskap AB Hatzenbichler Agro-Technik Kongskilde Kress Umweltschonende Landtechnik (K.U.L.T.) Maschio Gaspardo Schmotzer Steketee Thyregod
Camera - Image analysis	Inter- and intrarow cultivation	F. Poulsen Engineering Garford Farm Machinery Ltd. Steketee (Lemken)
Robot platform with a multi-sensor system	Inter- and/or intrarow weed control	NAÏO Technologies
Robot platform with a multi-sensor system	Inter-row weed cotnrol	FarmWise

Basically, all row crops can be treated with a camera-steered hoe, provided that the row spacing is large enough (>20 cm), and that the camera and software algorithm can recognize the crop rows. This mainly depends on the crop growth stage and weediness. For tractor-pulled implements, the detection of specific weed species for mechanical weeding is seen as redundant since all other plants except the crop can be considered weeds. For agricultural robots, however, detecting single weed plants may be necessary if individual weed plants are supposed to be targeted. For example, the removal of specific target plants where a zero threshold is enforced (e.g. *Rumex* spp. in pastures).

RTK-GPS can be a very accurate solution to guide an implement along crop rows because no crop-specific knowledge is required (Pérez-Ruiz *et al.*, 2012). Table 2.1.5-3 shows that RTK-GPS and camera vision have a similar accuracy in terms of locating crop plants or positioning cultivation tools. However, from the number of studies discussed in this review, it appears that camera-based systems for mechanical weeding seem to predominate. RTK-GPS requires a-priori knowledge (data) to be recorded, stored, and processed from seeding until harvest to locate each crop plant again so that a mechanical treatment can be applied. Real-time sensor systems like machine vision perform crop and weed plant identification instantly. This enables quick and efficient steering of the mechanical implements and even a curved crop row can be followed easily (Figure 2.1.5-2).

Figure 2.1.5-2: A camera-steered hoe follows curved rows of summer barley.

Table 2.1.5-3: Comparison of GNSS and machine vision guidance accuracy in different studies.

System	Task	Mean guidance accuracy error	Speed	Source
GPS	Seed location mapping	± 38 mm	0.83 m s^{-2}	(Ehsani, Upadhyaya and Mattson, 2004b)
	Mapping of sugar beet seeds while sowing	± 19 mm	0.56 m s^{-2}	(Griepentrog *et al.*, 2005b)
		± 31 mm	1.39 m/s^{-2}	
	Following a crop row and changing the transverse position of a cycloid hoe	± 16 mm	0.31 m s^{-2}	(Nørremark *et al.*, 2008)
		± 22 mm	0.52 m s^{-2}	
	Location mapping of transplanted row crops	± 20 mm	0.44 m s^{-2}	(Sun *et al.*, 2010)
	GPS-based control of a knife for intrarow weeding	± 6 mm	0.22 m s^{-2}	(Pérez-Ruiz *et al.*, 2012)
		± 9 mm	0.44 m s^{-2}	
	Precision mapping of transplanted tomato plants	± 32 mm	0.44 m s^{-2}	(Perez-Ruiz *et al.*, 2012)
Camera	Comparison of vision guided hoeing to manual steering	± 18 mm	1.8 m s^{-2}	(Home *et al.*, 2002)
		± 20 mm	3.1 m s^{-2}	
	Automatic guidance of a hoe in sugar beet	± 16 mm	1.67 m s^{-2}	(Tillett *et al.*, 2002)
	Evaluation of vision-guided hoeing in sugar beet	± 12 mm	1.4 m s^{-2}	(Wiltshire, Tillett and Hague, 2003)

Machine vision must also consider the changing crop plant height and width as the growing season advances. Of course, this increases the complexity of the software algorithm, but it also allows the precise guidance of cultivation tools near the crop rows or individual crop plants in every growth stage. Thus, crop damage is avoided to the best extent possible. Missing crop plants are also considered by camera-guided implements and the empty area is treated with the mechanical tools, whereas a machine which depends on GNSS references would keep the area of a missing crop plant untreated. A camera-steered side shifting frame for hoeing also has the advantage that the lateral re-positioning of the hoe is performed instantly and directly rather than indirectly through the position of the tractor when only GNSS is used (Fennimore *et al.*, 2016). However, large amounts of weeds can be a distraction for camera-steered systems (Slaughter *et al.*, 1999). Additional limitations for machine vision include shadow casting in bright sunlight, movement of the crop due to wind and large variations of the crop height and width within the field.

Mechanical weeding with machine vision can be facilitated by only identifying crop plants and to treat the entire area surrounding the crop plants regardless if any weeds are actually present (Guyer *et al.*, 1986). But making the additional effort to detect and identify weed species can become important if a low weed density with a high species diversity is desired to increase the biodiversity of a field. In such a scenario, an algorithm could be used to leave rare weed species unharmed. Increasing the diversity of weed species in an agroecosystem reduces the likelihood of a particular weed species occurring too frequently and having negative effects on the crop (Storkey *et al.*, 2018).

In the future, the technology of machine vision in mechanical weed control will continue to evolve and adapt to multiple cropping scenarios including very narrow row widths of less than 20 cm as can be found in conventional European cereal cropping systems where 12.5 or 15 cm are typical row spacings. The working widths of mechanical hoes are smaller than those of a modern sprayer for chemical weed control and therefore more time is required for mechanical weeding. However, camera steered hoeing implements lead to higher driving speeds and can at least partially compensate the time loss (Paarlberg *et al.*, 2013). This could make mechanical weeding feasible for conventional farmers as well. Another advantage of mechanical weed control is that it is not as sensitive to wind as spraying.

Laser and ultrasonic sensors on their own are not adequate sensor systems for mechanical weeding and are primarily assisting sensors for implement guidance. Especially in robotics, these are essential parts of the autonomous system where the machine is not guided by a human operator. The best use of those systems can, therefore, be seen in autonomous robots as additional guidance systems. Agricultural robots must rely on a combination of machine vision, GNSS, laser and ultra- sonic sensors (Bakker, Van Asselt, Bontsema and Van Henten, 2010; Young, Meyer and Woldt, 2014). Due to their elaborate setup, they face many difficulties. Already in the early 1990s, N. D. Tillett (1991) addressed this problem, stating that a completely autonomous system is difficult to realize because of certain tasks that a human operator must perform. Almost 30 years later, Merfield (2016) acknowledged the complexity of robotic systems but criticizes the current systems of still not being completely autonomous from a human operator. This may include repairs, route, and task planning as well as the adjustment of mechanical weeding tools. Further limitations include tasks deviating from the standard procedure. Peteinatos *et al.* (2014) argue that within certain conditions sensor-based systems can work effectively, but due to the ever-changing agricultural environment many difficulties arise. Therefore, completely autonomous systems, which are truly independent of a human operator, are not going to exist soon. On the other hand, the question arises how far the requirements for autonomous robots should strive and if completely autonomous systems are ever necessary and desired. Especially for mechanical weed control, a farmer's experience is crucial as soil and weather conditions play an important role in the timing of weed control strategies (Home *et al.*, 2002; Kunz *et al.*, 2015). The overall aim would be to have robots, which can decide on their own which tasks are necessary at a given time of the year. Collecting, organizing, and transferring the "agricultural knowledge" into an accessible database for robots is a demanding task on its own. Beforehand, robots must be developed to a point where they can make efficient use of such information.

Until now weed control is aiming at creating solutions for the current farming systems and with the logic that the system should achieve the best weed suppression with a minimum number of treatments (e.g. 1–3). Yet both prerequisites need to be examined again for the adaptation of robotic systems in weed management. For example, "Tertill" (Franklin Robotics, Billerica, USA) which is an autonomous robot for weed management in household gardens, might also be a solution in agriculture. Its concept of blindly repeating the same task every day, indiscriminately targeting all newly emerging plants can be an alternative solution. A swarm of simple, low-cost robots that can fulfill a simple task without specialization can be an option, especially in high-value row crops with a low weed tolerance. On the other hand, we might also reconsider our weed management approach. As traditional cultivations like vineyards have been modified to incorporate mechanization, we should also be acceptable in cultivation modifications that can facilitate simplified and effective weed control.

Apart from the previously mentioned implications, there are certain challenges to mechanical weeding that must be kept in mind now and in the future for commercial weeding implements. Many challenges do not arise from the combination of mechanical weeding and sensor guidance alone but rather from the nature of mechanical weeding itself. Mechanical weeding does not have a label stating the amount of necessary treatments like a typical herbicide. Even more so, the timing of the treatment application differs between regions and the local soil conditions. An advantage of sensor-guidance is the increased accuracy of the weeding operation. In the best-case scenario, this helps to increase the weed control efficacy and thus reduces the necessary amount of tractor passes with a mechanical implement. However, weeds are usually emerging in several waves throughout the growing season and the success of a weeding operation, therefore, always depends on the local weed pressure. Thus, there is no guarantee that a sensor-guided implement always leads to less treatment applications.

Future challenges are to design sensor-guided weeding implements with a high robustness and sustainability. This requires the exquisite support of the implement manufacturers. There must be an extensive network of support technicians throughout a country to guarantee that the machinery can be repaired and tended to by an expert without large time delays. Since mechanical weeding highly depends on the local weather conditions, long waiting periods for a machine to be up and running again could ruin an entire cropping season because weeds are not halting their growth. Also, in the effort to reduce the amount of micro-plastic in the environment it is advisable to use biodegradable rubber plastic for tools such as the finger weeders. Due to the high abrasion forces of the soil, such tools must be replaced regularly. Finding alternative solutions to petrol based plastic tools would contribute to a healthier environment in the future. Alternative fuel technology should also be addressed in agriculture. Harrowing or hoeing requires more fuel energy than spraying. Therefore, a new fuel technology must help to decrease the pollution by this type of field work. Possibly many small robots running on solar power are the better solution to future maintenance field work than a single tractor.

This study has shown that equipping machinery with sensors can help to increase weed control, though not all systems are feasible. Despite the benefits of sensor-equipped machinery, the downside can include higher repair and acquisition costs. Autonomous systems must also have a well-developed accident prevention system so that no damage is caused by self-operating machinery in the field or while traveling to the field along public roadways. When looking at the markets it is obvious that camera-steering systems are the main type of sensor used in mechanical weeding. However, the focus should not only rest on the direct weed control measures but rather the entire farming system is important to achieve acceptable weed control levels and high crop yields. Especially for mechanical weed control, the overall farm management must be focused on weed reduction. This includes indirect measures such as crop rotation, establishing suitable cover crops, and soil management (Kruidhof *et al.*, 2008) to create unfavorable growing conditions for weeds (Abbas *et al.*, 2018). Mechanical weeding, as a direct weed control measure, can only be the last step in the management cycle of a crop. Independent of the sensor system, the build-up of a strong weed plant community must be prevented. Otherwise, mechanical weeding is not going to produce a "clean" field. Therefore, mechanical weeding is the tip of the iceberg in the process of weed management irrespective of all the advancements made in sensor technology.

However, agricultural fields are a diverse environment and provide many challenges for a reliable implementation of sensors. Alternating weather conditions (Kurstjens and Kropff, 2001), different soil characteristics, a multitude of different cropping systems, plant debris, dust, water and the absence of controlled illumination are prime examples. Furthermore, the weeds themselves underlie a spatial heterogeneity because they are not uniformly distributed within fields (Cardina *et al.*, 1995; Clay *et al.*, 1999). Their appearance, color, and size vary (Gerhards *et al.*, 2002). Even the same weed plant can have different visual appearances during its different development stages. Site-specific weed management (SSWM) is part of the Precision Agriculture development and considers those spatial and temporal changes of weeds (Young *et al.*, 2014). In traditional plant protection, plants are treated as a community and not as single individuals. However, sensors enable us precise management of fields. Therefore, a shift towards treating even single crop plants has been made possible during the past years not only for chemical but also for mechanical weed control methods due to advanced sensor technology (Ehsani *et al.*, 2004; Griepentrog *et al.*, 2005a). The initial idea behind SSWM was to reduce the input of synthetic pesticides. For weeds, this can be achieved either by spot or patch spraying based on digital weed detection methods. However, the principles of site-specific weed control based on in-field data can now also be applied to mechanical weeding. This requires the adoption of weeding machines that can deal with a varying environment while pursuing the highest operation accuracy (Van Der Weide *et al.*, 2008). The benefits of the digitization in mechanical weed control include an increased precision which allows higher driving speeds and ultimately leads to higher ground coverage (Wilson, 2000; Home *et al.*, 2002). Thus, the efficacy of weed removal is increased (Grift *et al.*, 2008).

2.1.6 Conclusion

Mechanical weed control is a complex part of agriculture. It requires considerable experience of the farmer to develop a long-term concept to keep the weed pressure as low as possible. With the advancements of herbicide resistances and political efforts to minimize the use of pesticides, mechanical weeding and other alternative weed control methods must be developed further. The first smart cultivators that evolved from research were based on machine vision or RTK-GPS for inter-row hoeing. They are now widely used in many different crops. Especially wide spaced crops like sugar beets, maize or soybean are often hoed with these guidance systems. The benefits of such systems include a higher driving speed and increased precision which leads to a higher area coverage compared to manual steering. The development of robust machine vision algorithms continued and lead to commercial systems for individual crop plant treatments of the intrarow area. However, RTK-GPS proved to be too inaccurate for intrarow weeding and its use is restricted to inter-row guidance. However, the range of applications for GNSS guided systems is quite large. For example, autonomous vehicles rely on GNSS for the general navigation on a field and they can then perform seeding or mechanical weeding. Laser and ultrasonic sensor systems are mostly used as additional guidance sensors (e.g. obstacle avoidance) for autonomous robots. In general, a rise in the development of autonomous robots can be observed. Fleets of small robots are ideal to perform repetitive agricultural tasks and their application will increase in the future.

2.1.7 Declaration of Competing Interest

The authors declare that they have no known competing financial interests or personal relationships that could have appeared to influence the work reported in this paper.

2.1.8 References

ABBAS, T., ZAHIR, Z. A., NAVEED, M. AND KREMER, R. J. (2018) 'Limitations of Existing Weed Control Practices Necessitate Development of Alternative Techniques Based on Biological Approaches', in *Advances in Agronomy*. Academic Press, pp. 239–280.

ANDÚJAR, D., ESCOLÀ, A., ROSELL-POLO, J. R., FERNÁNDEZ-QUINTANILLA, C. AND DORADO, J. (2013) 'Potential of a terrestrial LiDAR-based system to characterise weed vegetation in maize crops', *Computers and Electronics in Agriculture*, **92**, pp. 11–15.

ANDÚJAR, D., WEIS, M. AND GERHARDS, R. (2012) 'An ultrasonic system for weed detection in cereal crops', *Sensors (Switzerland)*, **12**(12), pp. 17343–17357.

ASCARD, J. AND BELLINDER, R. R. B. (1996) 'Mechanical in-row cultivation in row crops', *Second International Weed Control Congress Copenhagen*, (1993), pp. 1121–1126.

ÅSTRAND, B. AND BAERVELDT, A. J. (2005) 'A vision based row-following system for agricultural field machinery', *Mechatronics*, **15**(2), pp. 251–269.

AUERNHAMMER, H. AND MUHR, T. (1991) 'GPS in a basic rule for environmental protection in agriculture'. American Society of Agricultural Engineers. Available at: http://agris.fao.org/agris-search/search.do?recordID=US9325817 (Accessed: 28 July 2019).

Autonomous weeding, agricultural robots - Naïo Technologies (2019). Available at: https://www.naio-technologies.com/en/ (Accessed: 2 August 2019).

BAERVELDT, A. AND ÅSTRAND, B. (2002) 'An agricultural mobile robot with vision-based perception for mechanical weed control', *Autonomous robots*, **13**, pp. 21–35.

BAKKER, T., ASSELT, K., BONTSEMA, J., MÜLLER, J. AND STRATEN, G. (2010) 'Systematic design of an autonomous platform for robotic weeding', *Journal of Terramechanics*, **47**(2), pp. 63–73.

BAKKER, T., VAN ASSELT, K., BONTSEMA, J. AND VAN HENTEN, E. J. (2010) 'Robotic weeding of a maize field based on navigation data of the tractor that performed the seeding', in *IFAC Proceedings Volumes (IFAC-PapersOnline)*. Available at: https://www.sciencedirect.com/science/article/pii/S1474667015310557 (Accessed: 28 July 2019).

BAKKER, T., VAN ASSELT, K., BONTSEMA, J., MÜLLER, J. AND VAN STRATEN, G. (2010) 'A path following algorithm for mobile robots', *Autonomous Robots*, **29**(1), pp. 85–97.

BAWDEN, O., KULK, J., RUSSELL, R., MCCOOL, C., ENGLISH, A., DAYOUB, F., LEHNERT, C. AND PEREZ, T. (2017) 'Robot for weed species plant-specific management', *Journal of Field Robotics*, **34**(6), pp. 1179–1199.

BECHAR, A. AND VIGNEAULT, C. (2017) 'Agricultural robots for field operations. Part 2: Operations and systems', *Biosystems Engineering*, **153**, pp. 110–128.

BILLINGSLEY, J. AND SCHOENFISCH, M. (1995) 'Vision-guidance of agricultural vehicles', *Autonomous Robots*. Kluwer Academic Publishers, **2**(1), pp. 65–76.

BILLINGSLEY, J. AND SCHOENFISCH, M. (1997) 'The successful development of a vision guidance system for agriculture', *Computers and Electronics in Agriculture*. Elsevier, **16**(2), pp. 147–163.

BOND, W., BURSTON, S., BEVAN, J. R. AND LENNARTSSON, M. E. (1998) 'Optimum Weed Removal Timing in Drilled Salad Onions and Transplanted Bulb Onions Grown in Organic and Conventional Systems', *Biological Agriculture and Horticulture*, **16**(2), pp. 191–201.

BORGELT, S. C., HARRISON, J. D., SUDDUTH, K. A. AND BIRRELL, S. J. (1996) 'Evaluation of GPS for applications in precision agriculture', *Applied Engineering in Agriculture*, **12**(6), pp. 633–638.

BOWMAN, G. AND OUTREACH, S. (2002) *Steel in the field: a farmer's guide to weed management tools*, *Sustainable Agriculture Network handbook series (USA)*.

BRANDSÆTER, L. O., MANGERUD, K. AND RASMUSSEN, J. (2012) 'Interactions between pre- and post-emergence weed harrowing in spring cereals', *Weed Research*, **52**(4), pp. 338–347.

DE BUCK, A. J., SCHOORLEMMER, H. B., WOSSINK, G. A. A. AND JANSSENS, S. R. M. (1999) 'Risks of post-emergence Weed control strategies IN Sugar beet: Development and application of a bio-economic model', *Agricultural Systems*, **59**(3), pp. 283–299.

BUSI, R., VILA-AIUB, M. M., BECKIE, H. J., GAINES, T. A., GOGGIN, D. E., KAUNDUN, S. S., LACOSTE, M., NEVE, P., NISSEN, S. J., NORSWORTHY, J. K., RENTON, M., SHANER, D. L., TRANEL, P. J., WRIGHT, T., YU, Q. AND POWLES, S. B. (2013) 'Herbicide-resistant weeds: From research and knowledge to future needs', *Evolutionary Applications*. John Wiley & Sons, Ltd, **6**(8), pp. 1218–1221.

CARDINA, J., SPARROW, D. H. AND MCCOY, E. L. (1995) 'Analysis of Spatial Distribution of Common Lambsquarters (Chenopodium album) in No-Till Soybean (Glycine max)', *Weed Science*, **43**(2), pp. 258–268.

Carré - Made for Agriculture (2020). Available at: http://www.carre.fr/en/crop-maintenance/weeding-robot/61-anatis.html (Accessed: 2 August 2019).

CHEN, L., KAEWKORN, S., HE, L., ZHANG, Q. AND KARKEE, M. (2016) 'Design and Evaluation of a Levelling System for a Weeding Robot', *IFAC-PapersOnLine*, **49**(16), pp. 299–304.

CIONI, F. AND MAINES, G. (2010) 'Weed Control in Sugarbeet', *Sugar Tech*, **12**(3–4), pp. 243–255.

'Claas CULTI CAM' (2020). Available at: http://www.claas-e-systems.com/en/oem-products/camera/.

CLAY, S. A., LEMS, G. J., CLAY, D. E., FORCELLA, F., ELLSBURY, M. M. AND CARLSON, C. G. (1999) 'Sampling weed spatial variability on a fieldwide scale', *Weed Science*, **47**(6), pp. 674–681.

CLOUTIER, D. C.; VAN DER WEIDE, R. Y., PERUZZI, A. AND LEBLANC, M. L. (2007) '8 Mechanical Weed Management', *Non-chemical Weed management*, p. 111.

COLAÇO, A. F., MOLIN, J. P., ROSELL-POLO, J. R. AND ESCOLÀ, A. (2018) 'Application of light detection and ranging and ultrasonic sensors to high-throughput phenotyping and precision horticulture: Current status and challenges', *Horticulture Research*, **5**(1).

CONESA-MUÑOZ, J., GONZALEZ-DE-SOTO, M., GONZALEZ-DE-SANTOS, P. AND RIBEIRO, A. (2015) 'Distributed multi-level supervision to effectively monitor the operations of a fleet of autonomous vehicles in agricultural tasks', *Sensors (Switzerland)*, **15**(3), pp. 5402–5428.

CONNOLLY, C. (2003) 'Vision guidance system facilitates high-speed inter-row weeding', *Industrial Robot*, **30**(5), pp. 410–413.

CORDILL, C. AND GRIFT, T. E. (2011) 'Design and testing of an intra-row mechanical weeding machine for corn', *Biosystems Engineering*, **110**(3), pp. 247–252.

DAY, W. (2011) 'Engineering advances for input reduction and systems management to meet the challenges of global food and farming futures', *Journal of Agricultural Science*, **149**(S1), pp. 55–61.

DRAYCOTT, A.P. (2008) *Development of sugar beet, Sugar beet*. Edited by A. Philip Draycott. Oxford, UK: Blackwell Publishing Ltd.

DUKE, S. O. (2012) 'Why have no new herbicide modes of action appeared in recent years?', *Pest Management Science*, pp. 505–512.

EDAN, Y. (1995) 'Design of an autonomous agricultural robot', *Applied Intelligence*. Kluwer Academic Publishers, **5**(1), pp. 41–50.

EHSANI, M. R., UPADHYAYA, S. K. AND MATTSON, M. L. (2004a) 'Seed location mapping using RTK GPS', *Transactions of the American Society of Agricultural Engineers*. American Society of Agricultural and Biological Engineers, **47**(3), pp. 909–914.

EHSANI, M. R., UPADHYAYA, S. K. AND MATTSON, M. L. (2004b) 'Seed location mapping using RTK GPS', *Transactions of the American Society of Agricultural Engineers*. American Society of Agricultural and Biological Engineers, **47**(3), pp. 909–914.

EMMI, L., GONZALEZ-DE-SOTO, M., PAJARES, G. AND GONZALEZ-DE-SANTOS, P. (2014) 'New trends in robotics for agriculture: Integration and assessment of a real fleet of robots', *The Scientific World Journal*, **2014**.

EUROPEAN GLOBAL NAVIGATION SATELLITE SYSTEMS AGENCY (2014) *European GNSS Agency*. Available at: https://www.gsa.europa.eu/ (Accessed: 4 July 2020).

VAN EVERT, F. K., SAMSOM, J., POLDER, G., VIJN, M., DOOREN, H. J. VAN, LAMAKER, A., VAN DER HEIJDEN, G. W. A. M., KEMPENAAR, C., VAN DER ZALM, T. AND LOTZ, L. A. P. (2011) 'A robot to detect and control broad-leaved dock (Rumex obtusifolius L.) in grassland', *Journal of Field Robotics*, **28**(2), pp. 264–277.

FarmDroid FD20 (2020). Available at: https://solar-andresen.com/farmdroid/ (Accessed: 30 June 2020).

FarmWise Robot (2020). Available at: https://farmwise.io/ (Accessed: 15 March 2020).

FENNIMORE, S. A., SLAUGHTER, D. C., SIEMENS, M. C., LEON, R. G. AND SABER, M. N. (2016) 'Technology for Automation of Weed Control in Specialty Crops', *Weed Technology*. Cambridge University Press (CUP), **30**(4), pp. 823–837.

FENNIMORE, S. A., SMITH, R. F., TOURTE, L., LESTRANGE, M. AND RACHUY, J. S. (2014) 'Evaluation and Economics of a Rotating Cultivator in Bok Choy, Celery, Lettuce, and Radicchio', *Weed Technology*, **28**(1), pp. 176–188.

FERNÁNDEZ-QUINTANILLA, C., PEÑA, J. M., ANDÚJAR, D., DORADO, J., RIBEIRO, A. AND LÓPEZ-GRANADOS, F. (2018) 'Is the current state of the art of weed monitoring suitable for site-specific weed management in arable crops?', *Weed Research*. Edited by R. Smith. John Wiley & Sons, Ltd (10.1111), pp. 259–272.

FINDLAY, S., CARREIRO, M., KRISCHIK, V. AND JONES, C. G. (1996) 'Effects of damage to living plants on leaf litter quality', *Ecological Applications*, **6**(1), pp. 269–275.

Franklin Robotics | Home of Tertill, the robotic garden weeder that is powered by the sun! (2020). Available at: https://www.franklinrobotics.com/ (Accessed: 28 July 2019).

Garford Farm Machinery Ltd | Precision Guided Hoes & InRow Weeders (2020). Available at: https://garford.com/ (Accessed: 7 April 2020).

GERHARDS, R., SÖKEFELD, M., TIMMERMANN, C., KÜHBAUCH, W. AND WILLIAMS, I. M. (2002) 'Site-specific weed control in maize, sugar beet, winter wheat, and winter barley', in *Precision Agriculture*, pp. 25–35.

GERRISH, J. B., FEHR, B. W., VAN EE, G. R. AND WELCH, D. P. (1997) 'Self-steering tractor guided by computer-vision', *Applied Engineering in Agriculture*, **13**(5), pp. 559–563.

GOBOR, Z. (2013) 'Mechatronic System for Mechanical Weed Control of the Intra-row Area in Row Crops', *KI - Künstliche Intelligenz*, **27**(4), pp. 379–383.

GONZALEZ-DE-SANTOS, P., RIBEIRO, A., FERNANDEZ-QUINTANILLA, C., LOPEZ-GRANADOS, F., BRANDSTOETTER, M., TOMIC, S., PEDRAZZI, S., PERUZZI, A., PAJARES, G., KAPLANIS, G., PEREZ-RUIZ, M., VALERO, C., DEL CERRO, J., VIERI, M., RABATEL, G. AND DEBILDE, B. (2017) 'Fleets of robots for environmentally-safe pest control in agriculture', *Precision Agriculture*, **18**(4), pp. 574–614.

GRIEPENTROG, H. W., CHRISTENSEN, S., SØGAARD, H. T., NØRREMARK, M., LUND, I. AND GRAGLIA, E. (2004) 'Robotic weeding', *EurAgEng04: Leuven, Belgium*, pp. 12–16.

GRIEPENTROG, H. W., NØRREMARK, M., NIELSEN, H. AND BLACKMORE, B. S. (2005a) 'Seed mapping of sugar beet', in *Precision Agriculture*. Kluwer Academic Publishers, pp. 157–165.

GRIEPENTROG, H. W., NØRREMARK, M., NIELSEN, H. AND BLACKMORE, B. S. (2005b) 'Seed mapping of sugar beet', in *Precision Agriculture*. Springer, pp. 157–165.

GRIFT, T., ZHANG, Q., KONDO, N. AND TING, K. (2008) 'A review of automation and robotics for the bio-industry', *Journal of Biomechatronics Engineering*, **1**(1), pp. 37–54.

GUNSOLUS, J. L. (1990) 'Mechanical and cultural weed control in corn and soybeans', *American Journal of Alternative Agriculture*, **5**(3), pp. 114–119.

GUYER, D. E., MILES, G. E., SCHREIBER, M. M., MITCHELL, O. R. AND VANDERBILT, V. C. (1986) 'Machine vision and image processing for plant identification', *Transactions of the ASAE*, **29**(6), pp. 1500–1507.

HANDS FREE HECTARE (2018) *Hands free hectare - Home*. Available at: http://www.handsfreehectare.com/ (Accessed: 2 August 2019).

HARKER, K. N. AND O'DONOVAN, J. T. (2013) 'Recent Weed Control, Weed Management, and Integrated Weed Management', *Weed Technology*, **27**(1), pp. 1–11.

HEAP, I. M. (2020) *The International Survey of Herbicide Resistant Weeds. Online. Internet.* Available at: www.weedscience.org (Accessed: 4 July 2020).

HEISEL, T., ANDREASEN, C. AND CHRISTENSEN, S. (2002) 'Sugarbeet yield response to competition from Sinapis arvensis or Lolium perenne growing at three different distances from the beet and removed at various times during early growth', *Weed Research*, **42**(5), pp. 406–413.

VAN HENTEN, E. J., VAN DIJK, G., KUYPERS, M. C., VAN TUIJL, B. A. J. AND VAN WILLIGENBURG, L. G. (2017) 'Motion Planning for a Cucumber Picking Robot', *IFAC Proceedings Volumes*, **33**(19), pp. 39–44.

HIREMATH, S. A., STEIN, A., TER BRAAK, C. J. F., VAN DER HEIJDEN, G. W. A. M. AND VAN EVERT, F. K. (2013) 'Laser range finder model for autonomous navigation of a robot in a maize field using a particle filter', *Computers and Electronics in Agriculture*, **100**, pp. 41–50.

HOME, M. C. W., TILLETT, N. D., HAGUE, T. AND GODWIN, R. J. (2002) 'An experimental study of lateral positional accuracy achieved during inter-row cultivation', in *Proceedings of the 5th EWRS Workshop on Physical and Cultural Weed Control. Scuola Superiore Sant'Anna di studi universitari e di perfezionamento, Pisa, Italy. 11-13 March 2002*, pp. 101–110.

KEICHER, R. AND SEUFERT, H. (2000) 'Automatic guidance for agricultural vehicles in Europe', *Computers and Electronics in Agriculture*, **25**(1–2), pp. 169–194.

KNEZEVIC, S. Z., EVANS, S. P., BLANKENSHIP, E. E., VAN ACKER, R. C. AND LINDQUIST, J. L. (2002) 'Critical period for weed control: the concept and data analysis', *Weed Science*. Page Press Publications, **50**(6), pp. 773–786.

KOLB, L. N. AND GALLANDT, E. R. (2012) 'Weed management in organic cereals: Advances and opportunities', *Organic Agriculture*, pp. 23–42.

KRUIDHOF, H. M., BASTIAANS, L. AND KROPFF, M. J. (2008) 'Ecological weed management by cover cropping: Effects on weed growth in autumn and weed establishment in spring', *Weed Research*, **48**(6), pp. 492–502.

KUNZ, C., WEBER, J. F., PETEINATOS, G. G., SÖKEFELD, M. AND GERHARDS, R. (2018) 'Camera steered mechanical weed control in sugar beet, maize and soybean', *Precision Agriculture*, **19**(4), pp. 708–720.

KUNZ, C., WEBER, J. AND GERHARDS, R. (2015) 'Benefits of Precision Farming Technologies for Mechanical Weed Control in Soybean and Sugar Beet—Comparison of Precision Hoeing with Conventional Mechanical Weed Control', *Agronomy*, **5**(2), pp. 130–142.

KUNZ, C., WEBER, J., JULIUS-KÜHN-ARCHIV, R. G.- AND 2016, U. (2016) 'Comparison of different mechanical weed control strategies in sugar beets', *Julius-Kühn-Archiv*, (452), p. 446.

KURSTJENS, D. A. G. AND KROPFF, M. J. (2001) 'The impact of uprooting and soil-covering on the effectiveness of weed harrowing', *Weed Research*, **41**(3), pp. 211–228.

Kverneland (2020). Available at: https://uk.kverneland.com/.

LANGSENKAMP, F., SELLMANN, F., KOHLBRECHER, M., KIELHORN, A., STROTHMANN, W., MICHAELS, A., RUCKELSHAUSEN, A. AND TRAUTZ, D. (2014) 'Tube Stamp for mechanical intra-row individual Plant Weed Control', *Agricultural Engineering International: the CIGR Ejournal*, pp. 1–11.

LATI, R. N., SIEMENS, M. C., RACHUY, J. S. AND FENNIMORE, S. A. (2016) 'Intrarow Weed Removal in Broccoli and Transplanted Lettuce with an Intelligent Cultivator', *Weed Technology*, **30**(3), pp. 655–663.

VAN DER LINDEN, S., MOUAZEN, A. M., ANTHONIS, J., RAMON, H. AND SAEYS, W. (2008) 'Infrared laser sensor for depth measurement to improve depth control in intra-row mechanical weeding', *Biosystems Engineering*, **100**(3), pp. 309–320.

LÖTJÖNEN, T. AND MIKKOLA, H. (2000) 'Three mechanical weed control techniques in spring cereals', *Agricultural and Food Science in Finland*, **9**(4), pp. 269–278.

MACHLEB, J., KOLLENDA, B. L., PETEINATOS, G. G. AND GERHARDS, R. (2018) 'Adjustment of weed hoeing to narrowly spaced cereals', *Agriculture (Switzerland)*, **8**(4).

MELANDER, B., CIRUJEDA, A. AND JØRGENSEN, M. H. (2003) 'Effects of inter-row hoeing and fertilizer placement on weed growth and yield of winter wheat', *Weed Research*. John Wiley & Sons, Ltd, **43**(6), pp. 428–438.

MELANDER, B. AND HARTVIG, P. (1997) 'Yield responses of weed-free seeded onions [Allium cepa (L.)] to hoeing close to the row', *Crop Protection*, **16**(7), pp. 687–691.

MELANDER, B., LATTANZI, B. AND PANNACCI, E. (2015) 'Intelligent versus non-intelligent mechanical intra-row weed control in transplanted onion and cabbage', *Crop Protection*, **72**, pp. 1–8.

MELANDER, B., RASMUSSEN, I. A. AND BÀRBERI, P. (2005) 'Integrating physical and cultural methods of weed control— examples from European research', *Weed Science*. Cambridge University Press (CUP), **53**(3), pp. 369–381.

MELANDER, B., RASMUSSEN, I. A. AND BÀRBERI, P. (2006) 'Integrating physical and cultural methods of weed control— examples from European research', *Weed Science*, **53**(3), pp. 369–381.

MERFIELD, C. N. (2016) 'Robotic weeding's false dawn? Ten requirements for fully autonomous mechanical weed management', *Weed Research*. Edited by C. Kempenaar, **56**(5), pp. 340–344.

MIDTIBY, H. S., ÅSTRAND, B., JØRGENSEN, O. AND JØRGENSEN, R. N. (2016) 'Upper limit for context–based crop classification in robotic weeding applications', *Biosystems Engineering*, **146**, pp. 183–192.

MONTALVO, M., GUERRERO, J., ROMEO, J., EMMI, L., GUIJARRO, M. AND PAJARES, G. (2013) 'Automatic expert system for weeds/crops identification in images from maize fields', *Expert Systems with Applications*, **40**(1), pp. 75–82.

MÜTER, M., DAMEROW, L. AND LAMMERS, P. S. (2014) *Kameragesteuerte mechanische Unkrautbekämpfung in Pflanzenreihen, Landtechnik*. Available at: http://195.37.233.11/landtechnik/article/view/2014-69-3-120-124 (Accessed: 28 July 2019).

NOF, S. Y. (2009) *Springer handbook of automation*. Edited by S. Y. Nof. Springer Berlin Heidelberg.

NOGUCHI, N., WILL, J., REID, J. AND ZHANG, Q. (2004) 'Development of a master-slave robot system for farm operations', *Computers and Electronics in Agriculture*, **44**(1), pp. 1–19.

NØRREMARK, M., GRIEPENTROG, H. W., NIELSEN, J. AND SØGAARD, H. T. (2008) 'The development and assessment of the accuracy of an autonomous GPS-based system for intra-row mechanical weed control in row crops', *Biosystems Engineering*, **101**(4), pp. 396–410.

NØRREMARK, M., GRIEPENTROG, H. W., NIELSEN, J. AND SØGAARD, H. T. (2012) 'Evaluation of an autonomous GPS-based system for intra-row weed control by assessing the tilled area', *Precision Agriculture*, **13**(2), pp. 149–162.

ORIADE, C. AND FORCELLA, F. (1999) 'Maximizing efficacy and economics of mechanical weed control in row crops through forecasts of weed emergence', *Journal of Crop Production*, **2**(1), pp. 189–205.

PAARLBERG, K. R., HANNA, H. M., ERBACH, D. C. AND HARTZLER, R. G. (2013) 'Cultivator design for interrow weed control in no-till corn', *Applied Engineering in Agriculture*, **14**(4), pp. 353–361.

PÉREZ-RUIZ, M., GONZALEZ-DE-SANTOS, P., RIBEIRO, A., FERNÁNDEZ-QUINTANILLA, C., PERUZZI, A., VIERI, M. AND AGÜERA, J. (2015) 'Highlights and preliminary results for autonomous crop protection', *Computers and Electronics in Agriculture*, **110**, pp. 150–161.

PÉREZ-RUIZ, M., SLAUGHTER, D. C., GLIEVER, C. J. AND UPADHYAYA, S. K. (2012) 'Automatic GPS-based intra-row weed knife control system for transplanted row crops', *Computers and Electronics in Agriculture*. Elsevier, pp. 41–49.

PEREZ-RUIZ, M., SLAUGHTER, D. C., GLIEVER, C. AND UPADHYAYA, S. K. (2012) 'Tractor-based Real-time Kinematic-Global Positioning System (RTK-GPS) guidance system for geospatial mapping of row crop transplant', *Biosystems Engineering*. Academic Press, **111**(1), pp. 64–71.

PÉREZ-RUIZ, M. AND UPADHYAYA, S. (2012) 'GNSS in Precision Agricultural Operations', in *New Approach of Indoor and Outdoor Localization Systems*. InTech.

PETEINATOS, G. G., WEIS, M., ANDÚJAR, D., RUEDA AYALA, V. AND GERHARDS, R. (2014) 'Potential use of ground-based sensor technologies for weed detection', *Pest Management Science*, pp. 190–199.

Platform PUMAgri - Sitia - Bancs d'Essais et Innovation Robotique (2020). Available at: http://www.sitia.fr/en/solution-innovation-en/systems-innovative/platform-pumagri/ (Accessed: 2 August 2019).

RASMUSSEN, I. A. (2004) 'The effect of sowing date, stale seedbed, row width and mechanical weed control on weeds and yields of organic winter wheat', *Weed Research*, **44**(1), pp. 12–20.

RASMUSSEN, J., BIBBY, B. M. AND SCHOU, A. P. (2008) 'Investigating the selectivity of weed harrowing with new methods', *Weed Research*. John Wiley & Sons, Ltd (10.1111), **48**(6), pp. 523–532.

RASMUSSEN, J., GRIEPENTROG, H. W., NIELSEN, J. AND HENRIKSEN, C. B. (2012) 'Automated intelligent rotor tine cultivation and punch planting to improve the selectivity of mechanical intra-row weed control', *Weed Research*, **52**(4), pp. 327–337.

RASMUSSEN, J., MATHIASEN, H. AND BIBBY, B. M. (2010) 'Timing of post-emergence weed harrowing', *Weed Research*, **50**(5), pp. 436–446.

REID, J. F. AND SEARCY, S. W. (1988) 'An Algorithm for Separating Guidance Information from Row Crop Images', *Transactions of the ASAE*, **31**(6), pp. 1624–1632.

Reihenführung - Schmotzer Hacktechnik (2020). Available at: https://www.schmotzer-ht.de/portfolio/reihenfuhrung/ (Accessed: 7 April 2020).

REISER, D., VÁZQUEZ-ARELLANO, M., PARAFOROS, D. S., MARRIDO-IZARD, M. AND GRIEPENTROG, H. W. (2018) 'Iterative individual plant clustering in maize with assembled 2D LiDAR data', *Computers in Industry*, **99**, pp. 42–52.

RIEMENS, M. M., VAN DER WEIDE, R. Y., BLEEKER, P. O. AND LOTZ, L. A. P. (2007) 'Effect of stale seedbed preparations and subsequent weed control in lettuce (cv. Iceboll) on weed densities', *Weed Research*. John Wiley & Sons, Ltd (10.1111), **47**(2), pp. 149–156.

Robozero Weeding Robot (2020). Available at: http://en.energreen.it/robo-remote-controlled-machines/robozero-remote-controlled-wheeled-cutting-machines/ (Accessed: 7 April 2020).

RUCKELSHAUSEN, A., KLOSE, R., LINZ, A., MARQUERING, J., THIEL, M. AND TÖLKE, S. (2006) 'Autonome roboter zur unkrautbekämpfung', in *Journal of Plant Diseases and Proctectio, Supplement*, pp. 173–180.

RUEDA-AYALA, V., PETEINATOS, G., GERHARDS, R. AND ANDÚJAR, D. (2015) 'A non-chemical system for online weed control', *Sensors*, **15**(4), pp. 7691–7707.

SLAUGHTER, D., CHEN, P., AGRICULTURE, R. C.-P. AND 1999, U. (1999) 'Vision Guided Precision Cultivation', *Precision Agriculture*, **1**(2), pp. 199–216.

SLAUGHTER, D., GILES, D., IN, D. D.-C. AND ELECTRONICS AND 2008, U. (2008) 'Autonomous robotic weed control systems: A review', *Computers and Electronics in Agriculture*, **61**(1), pp. 63–78.

STORKEY, J. AND NEVE, P. (2018) 'What good is weed diversity?', *Weed Research*. Edited by M. Liebman. Blackwell Publishing Ltd, **58**(4), pp. 239–243.

SUN, H., SLAUGHTER, D. C., RUIZ, M. P., GLIEVER, C., UPADHYAYA, S. K. AND SMITH, R. F. (2010) 'RTK GPS mapping of transplanted row crops', *Computers and Electronics in Agriculture*. Elsevier, **71**(1), pp. 32–37.

TILLETT, N. D. (1991) 'Automatic guidance sensors for agricultural field machines:A review', *Journal of Agricultural Engineering Research*, **50**(C), pp. 167–187.

TILLETT, N., HAGUE, T., AGRICULTURE, S. M.-C. AND ELECTRONICS IN AND 2002, U. (2002) 'Inter-row vision guidance for mechanical weed control in sugar beet', *Computers and Electronics in Agriculture*, **33**(3), pp. 163–177.

TILLETT, N., HAGUE, T., GRUNDY, A., ENGINEERING, A. D.-B. AND 2008, U. (2008) 'Mechanical within-row weed control for transplanted crops using computer vision', *Biosystems Engineering*, **99**(2), pp. 171–178.

TILLETT, N., RESEARCH, T. H.-J. OF A. E. AND 1999, U. (1999) 'Computer-vision-based hoe guidance for cereals - An initial trial', *Journal of Agricultural and Engineering Research*, **74**(3), pp. 225–236.

TYLER, D. A., ROBERT, P. C., RUST, R. H. AND LARSON, W. E. (1993) 'Positioning Technology (GPS)', in *Proceedings of Soil Specific Crop Management*. American Society of Agronomy, Crop Science Society of America, Soil Science Society of America, pp. 159–165.

VANGESSEL, M. J., SCHWEIZER, E. E., WILSON, R. G., WILES, L. J. AND WESTRA, P. (1998) 'Impact of Timing and Frequency of In-Row Cultivation for Weed Control in Dry Bean (Phaseolus vulgaris)', *Weed Technology*, **12**(3), pp. 548–553.

VAN DER WEIDE, R. Y., BLEEKER, P. O., ACHTEN, V. T. J. M., LOTZ, L. A. P., FOGELBERG, F. AND MELANDER, B. (2008) 'Innovation in mechanical weed control in crop rows', *Weed Research*, pp. 215–224.

WELSH, J. P., BULSON, H. A. J., STOPES, C. E., FROUD-WILLIAMS, R. J. AND MURDOCH, A. J. (1999) 'The critical weed-free period in organically-grown winter wheat', *Annals of Applied Biology*. Association of Applied Biologists, **134**(3), pp. 315–320.

WILSON, J. N. (2000) 'Guidance of agricultural vehicles - A historical perspective', *Computers and Electronics in Agriculture*, **25**(1–2), pp. 3–9.

WILTSHIRE, J. J. J., TILLETT, N. D. AND HAGUE, T. (2003) 'Agronomic evaluation of precise mechanical hoeing and chemical weed control in sugar beet', *Weed Research*, **43**(4), pp. 236–244.

YOUNG, S. L., MEYER, G. E. AND WOLDT, W. E. (2014) 'Future directions for automated weed management in precision agriculture', in *Automation: The Future of Weed Control in Cropping Systems*. Dordrecht: Springer Netherlands, pp. 249–259.

ZHANG, N., WANG, M. AND WANG, N. (2002) 'Precision agriculture—a worldwide overview', *Computers and electronics in agriculture*, **36**(2–3), pp. 113–132.

ZIMDAHL, R. (2004) *Weed-crop competition: A Review*. 2nd editio. Ames: Blackwell Publishing Ltd.

2.2 Adjustment of weed hoeing to narrowly spaced cereals

Jannis Machleb*, Benjamin L. Kollenda, Gerassimos G. Peteinatos and Roland Gerhards

Institute of Phytomedicine, Department of Weed Science, University of Hohenheim, 70599 Stuttgart, Germany; benjamin.kollenda@uni-hohenheim.de (B.L.K.); G.Peteinatos@uni-hohenheim.de (G.G.P.); Roland.Gerhards@uni-hohenheim.de (R.G.)

* Correspondence: jmachleb@uni-hohenheim.de; Tel.: +49-711-459-22394

Published in: *Agriculture* (2018), 8(4), 54.

The original publication is available at:
https://doi.org/10.3390/agriculture8040054

2.2.1 Abstract

Weed hoeing can be successfully performed in wide row crops, such as sugar beet, maize, soybean and wide spaced cereals. However, little experience is available for hoeing in narrow cereal row spaces below 200 mm. Yet, mechanical weed control can pose an alternative to herbicide applications by reducing the herbicide resistant populations present in the field. In this experiment, it was investigated whether hoeing is feasible in cereals with 150 and 125 mm row spacings. The trial was set up at two locations (Ihinger Hof and Kleinhohenheim) in southwest Germany. Three different conventional hoeing sweeps, a goosefoot sweep, a no-till sweep and a down-cut side knife were adjusted to the small row widths, and hoeing was performed once with a tractor and a standard hoeing frame which was guided by a second human operator. The average grain yield, crop and weed biomass, and weed control efficacy of each treatment were recorded. The goosefoot and no-till sweep were tested at driving speeds of 4 and 6 km h^{-1}. The down-cut side knife was applied at 4 km h^{-1}. The results indicate that hoeing caused no yield decrease in comparison to a conventional herbicide application or manual weeding. The highest yield with a mechanical treatment was recorded for the no-till sweeps at both trial locations. Hoeing was performed successfully in narrowly spaced cereals of 150 and 125 mm, and the weed control efficacy of the mechanical treatments ranged from 50.9% at Kleinhohenheim to 89.1% at Ihinger Hof. Future experiments are going to focus on more distinct driving speeds ranging from 2 to 10 km h^{-1} and performing more than one pass with the hoe. Additionally, combining the mechanical weeding tools with a camera-steered hoeing frame could increase accuracy, allow for higher working speeds and substitute the second human operator guiding the hoe.

Keywords: mechanical; hoeing; control; weeds; cereals; narrow; yield

2.2.2 Introduction

Mechanical weed control can be a successful tool for integrated pest management, in order to suppress herbicide resistant weeds. In cereals, a harrow is the most common tool for mechanical weed control of the inter- and intrarow spaces (Rasmussen, 1992). It can mainly target small weed seedlings which are uprooted or covered with soil by the metal tines. Larger weeds, perennial species and monocot weed species, in particular, are less affected by the harrowing. Due to these limitations, the interrow spaces between crop rows could be treated with more aggressive methods, such as hoeing (Terpstra and Kouwenhoven, 1981; Tillett and Hague, 1999; Melander, Cirujeda and Jørgensen, 2003). Interrow hoeing is practiced in organic cereal farming, with row distances of at least 200 mm in Northern Europe. According to Boström, Anderson and Wallenhammar (2012), doubling the row space from 120 mm to 240 mm, however, reduces grain yields in cereals by 12–16%. These findings are supported by Fahad *et al.* (2015), who stated that narrow row spacings lead to an increase in yield compared to wide row spacings. Moreover, narrowly seeded cereals can better suppress weed infestation compared to wider sowing (Mohler, 2001; Kolb, Gallandt and Mallory, 2012; Gallandt *et al.*, 2015). This emphasizes the need to develop new hoeing systems for narrowly spaced cereal systems (Melander, Cirujeda and Jørgensen, 2003). Row widths below 200 mm demand adjusted weeding blades and steering technologies to fit into the narrow row spaces and to guide the blades between the crop rows (Tillett and Hague, 2006). Camera-steered systems are already available on the market for wide row crops, such as sugar beet, maize and soybeans (Tillett *et al.*, 2003; Wiltshire, Tillett and Hague, 2003; Tillett and Hague, 2006; Kunz, Weber and Gerhards, 2015; Kunz *et al.*, 2017). Studies with vision-based hoe guidance have also been undertaken in wide cereal row spacings (Tillett and Hague, 1999, 2006; Home *et al.*, 2002; Tillett *et al.*, 2003). Yet, before adjusting sensor systems to narrow cereal rows, the impact of hoeing on crop yield, crop and weed biomass as well as on the weed control efficacy (WC) have to be investigated.

At two locations, three different cultivation sweeps were adjusted to row spacings of 150 and 125 mm, and spring barley and spring oats were tested at different driving speeds. The aim of this study was to assess grain yield, aboveground crop and weed biomass as well as the weed control efficacy for each tool. Untreated control plots, herbicide application and manual weed control served as references to the mechanical treatments.

2.2.3 Materials and Methods

2.2.3.1 Experimental Sites and Design

In 2017, two field trials were conducted at separate research locations of the University of Hohenheim in the southwest of Germany. The aim was to test the feasibility and performance of three different mechanical weeding tools in narrow row distances of 150 and 125 mm in spring cereals. The first location, Ihinger Hof (IHO, 48.74° N, 8.92° E), was a conventional farming system, situated near Renningen, at an altitude of 478 m above sea level. The second location was at the organically certified research site of Kleinhohenheim (KH, 48.73° N, 9.20° E) with an elevation of 400 m above sea level. The average annual rainfall for both research locations was similar with 690 mm (IHO) and 700 mm (KH). On both farms, the soil type was a loamy clay. The top 300 mm of soil was composed as follows: at IHO, there was 27.92% clay, 12.49% sand and 59.59% silt. At KH, the composition was 23.35% clay, 11.74% sand and 64.91% silt. Therefore, similar soil conditions existed at both locations. Due to the loam it was important to have dry conditions for the mechanical applications, otherwise no proper soil mixing effect would have been achieved.

The trials were set up as a randomized complete block design with four repetitions. The trial at IHO consisted of seven treatments. Treatments included an untreated control (CON), a single herbicide application (HERB) and five mechanical treatments, namely goosefoot sweeps at 4 km h^{-1} (GFS(4)), goosefoot sweeps at 6 km h^{-1} (GFS(6)), no-till sweeps at 4 km h^{-1} (NTS(4)), no-till sweeps at 6 km h^{-1} (NTS(6)) and down-cut side knives at 4 km h^{-1} (DSK). A detailed treatment description can be found in Table 2.2.7-1. Spring barley (cv. "Planet") was sown on 29 March 2017 with 300 viable seeds m^{-2} at a row distance of 150 mm. Mechanical treatments were applied once at crop stage BBCH (Biologische Bundesanstalt, Bundessortenamt und CHemische Industrie) 14–15 (Lancashire *et al.*, 1991). The weed development ranged between the cotyledon stage and the four true leaf stage. The herbicide "Axial komplett" (45 g active ingredient (a.i.) L^{-1} pinoxaden + 5 g a.i. L^{-1} florasulam) was sprayed once with 1.1 L ha^{-1} at crop stage BBCH 14 with a plot sprayer (Schachtner-Fahrzeug- und Gerätetechnik, Ludwigsburg, Germany) at 2.5 km h^{-1} and a calibration to deliver 200 L ha^{-1}. The untreated control was left untreated for the entire growing season. The plot size at IHO was 3 m by 12 m. In order to reduce possible border effects, only the 10 center rows of each plot were harvested. At IHO, the hoe was mounted on a Fendt tractor (model 207) with 2 m wide tracks.

Similar to IHO, the experiment at KH consisted of seven treatments: untreated control (CON), manual weeding (MANW), GFS(4), GFS(6), NTS(4), NTS(6) and DSK. A detailed treatment description can be found in Table 2.2.7-1. Due to the organic farming system, the herbicide application had to be replaced by two times manual weeding at crop stages BBCH 17–18 and BBCH 30. Manual weeding was performed by walking along the tractor tracks in order to avoid crop plant damage. Weeds were pulled by hand from the soil without the use of a garden hoe. Therefore, soil disturbance was reduced to a minimum. The first manual weed removal at BBCH 17–18 was conducted at the same time as the application of the mechanical treatments and again at 1 week after the herbicide application which was performed at IHO (BBCH 14). Spring oats cv. "Max" was sown on 17 March 2017 with 350 viable seeds m^{-2} and a row distance of 125 mm. Mechanical treatments were applied at crop stage BBCH 17–18. At KH, the weed plant size was between the first and fifth true leaf during the application of the mechanical treatments. The plot size was 1.5 m by 12 m. Again, the harvesting of grains was performed in the 10 center rows only to reduce possible border effects. The working width of the hoe was adjusted to the plot size of 1.5 m and a Fendt tractor (model GT 350) with 1.5 m wide tracks was used for hoeing at Kleinhohenheim.

2.2.3.2 Weeding tool description and implementation

Mechanical weeding tools were taken from different areas of agriculture, like vegetable farming, and adjusted for cereal cultivation of conventional farms in Northern Europe where typical row spacings are 150 mm and 125 mm. As with all mechanical weeding tools, adjustment of the hoeing depth, forward travel speed and tool width are essential to avoid crop damage or unsatisfying weed control (Kouwenhoven and Terpstra, 1979; Terpstra and Kouwenhoven, 1981). Lowering the sweeps too deep into the ground may create large lumps of soil in which weeds can continue to grow (Mattsson, Nylander and Ascard, 1990). In order to avoid crop damage due to misalignment of the sweeps, they were adjusted to their respective row width outside of the experiment in separate plots at each location. These additional plots consisted of the same crop with the same row distance as the actual experiment. This ensured optimal results when hoeing inside the experimental plots and minimized the risk of destroying plots. Other factors influencing the weeding result are the type and moisture content of the soil (Bàrberi, 2002). Favorable conditions for hoeing, namely dry and sunny weather with loose topsoil, were awaited to ensure optimal results. Hoeing was performed once at BBCH 14–15 in spring barley and at BBCH 17–18 in spring oats with a standard hoeing frame (Argus, K.U.L.T. Kress Umweltschonende Landtechnik GmbH, Vaihingen an der Enz, Germany). The working width was set to 3 m and 1.5 m at IHO and KH, respectively. The frame was equipped with K.U.L.T. DUO-parallelograms, as depicted in Figure 2.2.3-1. One parallelogram treated two interrow spaces. The DUO-parallelograms were chosen because they meet the requirements of a delicate tool for narrow row spaces while providing sufficient stability for effective mechanical weed control. A consistent soil working depth was achieved by adjusting the wheel height of each parallelogram, and uniform soil pressure was applied with a spring on every parallelogram. The sweeps were mounted behind the parallelograms on rigid toolbars, which were adjustable to the different hoeing widths at IHO and KH.

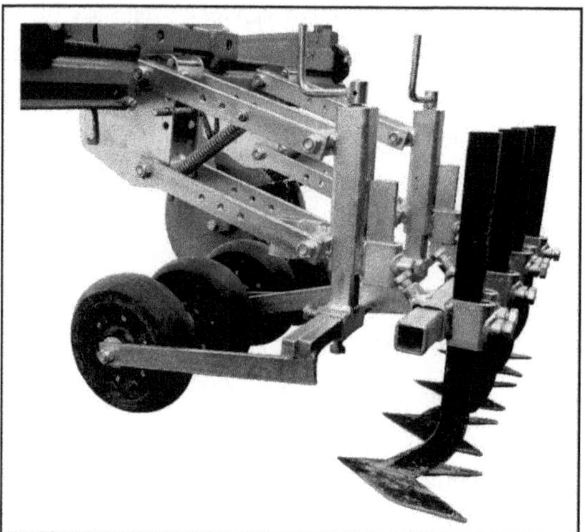

Figure 2.2.3-1: Overview of the parallelogram system: two K.U.L.T. DUO-parallelograms with no-till sweeps for 150 mm row spacing.

None of the sweeps on the market are small enough to enable hoeing in small row spaces. Hence, for each sweep, a specific amount of blade width was removed either on both sides (e.g., GFS and NTS) or in the case of DSK, on one side only. Cutting was performed with an angle grinder in a straight line from the front until the back of the blades. The cutting edges were smoothed and sharpened to ensure good soil penetration. Figure 2.2.3-2 depicts the three sweep types used in this study. GFS and NTS were centered inside the interrow area, leaving a safety margin of >20 mm from the crop plants on each side. Both DSK were placed with the vertical knife as close to the crop as possible, leaving a 20 mm safety margin from the crop plants on the left and right sides. After the successful adjustment of each tool, the field application was performed. The hoe was manually steered by a second human operator riding in the back of the tractor aligning the frame with the crop rows.

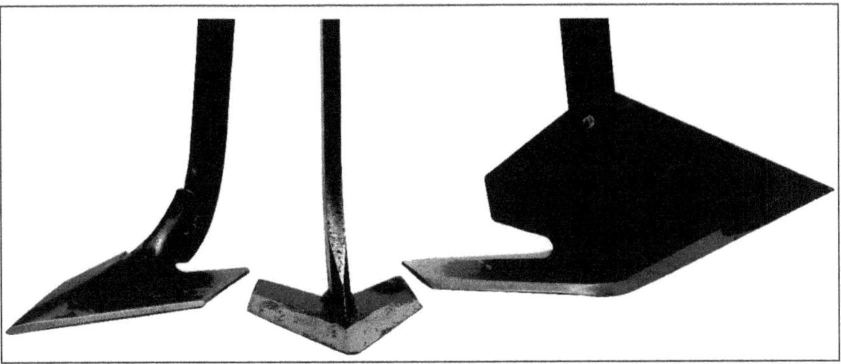

Figure 2.2.3-2: Group image of the three cultivation sweeps used in this study. From left to right: goosefoot sweep, no-till sweep and down-cut side knife. The depicted sweeps were adjusted to a row spacing of 150 mm.

2.2.3.2.1 Goosefoot Sweeps

Goosefoot sweeps are a common type of interrow weeding tool. The suggested working depth is between 20 to 30 mm. The commercial GFS were 160 mm in width. A safety margin of minimum 20 mm towards the cereal row on each blade side was taken into account to prevent crop damage. Cutting away 30 mm on each side resulted in a goosefoot sweep-width of 100 mm for the 150 mm row spacing in spring barley. For the smaller row spacing of 125 mm, a GFS-width of 80 mm remained. A visualization of the amount of sweep width taken away can be found in Table 2.2.3-1. The angle of the original sweeps or other features were not altered. The goosefoot sweeps are concave with a 20° drop-off, referred to as the sweep angle (Welsh *et al.*, 2002), from the center to the outer points of the blade edges. They have a high "crown" with a moderate "shoulder" (Bowman and Outreach, 2002). While traveling forward in the soil, the arched shape of the GFS leads to increased turbulences of soil particles. This causes the soil to be thrown sideways even at slower speeds when compared to other sweep types. The working depth of each goosefoot sweep was set to 20 mm. The rake angle is defined as the angle of the sweep in relation to the horizontal ground (Welsh *et al.*, 2002). It can be altered by adjusting the angle of the hoeing frame and was set to 5° for the GFS in order to allow for good soil penetration. One goosefoot sweep per interrow space was mounted onto the hoe. The GFS were tested at tractor driving speeds of 4 and 6 km h^{-1}.

Table 2.2.3-1: Top view of the commercial and adjusted goosefoot sweeps with the narrow row widths of 150 and 125 mm at Ihinger Hof and Kleinhohenheim, respectively. Hatched lines symbolize the area cut off from the commercial 160 mm wide goosefoot sweep. The plants to the left and right of the sweeps symbolize cereal rows.

	Ihinger Hof	**Kleinhohenheim**
	Row Width 150 mm	**Row Width 125 mm**
Commercial blade width 160 mm 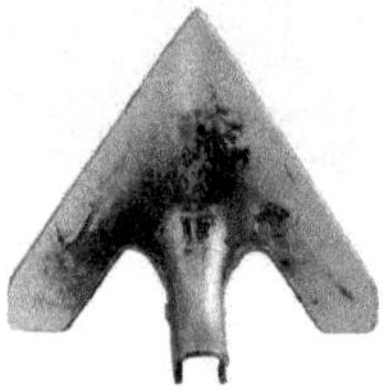	Adjusted blade width 100 mm 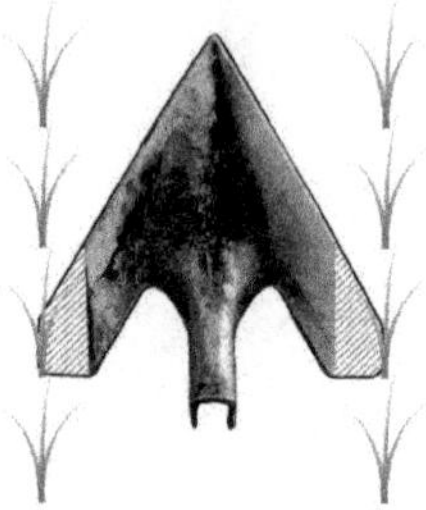	Adjusted blade width 80 mm 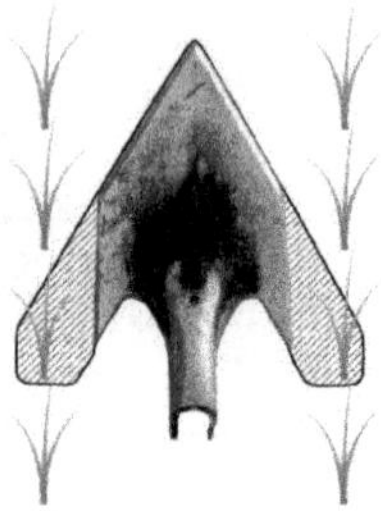

2.2.3.2.2 No-Till Sweeps

In contrast to the goosefoot sweeps, the no-till sweeps have a flat shape, without any angle of elevation (see Table 2.2.3-2 and Figure 2.2.3-2 for a direct comparison). Therefore, their sweep angle can be defined as 0°. Their crown is flat with low shoulders (Bowman and Outreach, 2002). NTS work, similar to GFS, in the center of the interrow space. However, less soil is disturbed compared to the GFS; tall weeds can be cut, and small weeds are uprooted. Due to their flat shape, the soil-burial effect is less than with GFS. Thus, a NTS moves less soil towards the crop row than a GFS with the same dimensions, working at the same driving speed. The NTS were also cut to widths of 100 mm (150 mm row spacing) and 80 mm (125 mm row spacing). It was seen as necessary to test both designs, GFS and NTS, for a comparison of their effects on grain yield, crop and weed biomass as well as on the weed control efficacy. Again, one NTS was mounted per interrow space. The working depth was adjusted to 20 mm and the tractor driving speeds were 4 and 6 km h^{-1}. Again, a rake angle of 5° was anticipated to ensure soil penetration without scrubbing of the sweeps over the soil surface.

Table 2.2.3-2: Top view of the commercial and the adjusted no-till sweeps to the narrow row widths of 150 and 125 mm at Ihinger Hof and Kleinhohenheim, respectively. Hatched lines symbolize the area cut off from the commercial 160 mm wide no-till sweep. The plants left and right of the sweeps symbolize cereal rows.

	Ihinger Hof	**Kleinhohenheim**
	Row Width 150 mm	**Row Width 125 mm**
Commercial blade width 160 mm	Adjusted blade width 100 mm	Adjusted blade width 80 mm

2.2.3.2.3 Down-Cut Side Knives

The down-cut side knife originally stems from vegetable cultivation systems. It consists of a blade with a 90° bend which ends in a knife-like blade cutting through the soil horizontally. Therefore, DSK have a blade sweep angle of 0°, similar to the principle of NTS. However, a rake angle of 5° had to be applied because otherwise, the DSK would not have penetrated the soil at IHO and KH. The vertical part of the blade, which is above ground, was placed near the crop row, moving parallel to the row, in the region between the inter- and intrarow space. Its main purpose is to precut the dead plant matter that is crossing the blades' path. Due to this design, no soil is moved into the intrarow space of the crop rows. Since each DSK consists of only one vertical blade, two are needed to cover each row. The side knives were cut to 90 mm (150 mm row spacing) and 70 mm (125 mm row spacing) in width. A detailed view of the DSK is depicted in Table 2.2.3-3. The first DSK was placed with the vertical part on the right border between the inter- and intrarow area in the direction of movement. Thus, the blade of the DSK was left-oriented and moved over the majority of the interrow space. The second DSK can be presented as a symmetrical mirror image of the first one. The vertical part was placed on the left border between the inter- and intrarow area and the knife blade was respectively right-oriented. The two blades of the side knives had 33% and 43% overlap in the middle of the interrow space, for the row distances of 150 mm and 125 mm, respectively. The working depth for the DSK was also set to 20 mm.

Table 2.2.3-3: Side view of the commercial down-cut side knife and top view of the down-cut side knives adjusted to the narrow row widths of 150 and 125 mm at Ihinger Hof and Kleinhohenheim, respectively. Hatched lines (here only shown for the 125 mm row width) indicate the blade length that was cut off from the commercial knife blade to fit the side knives between the cereal rows. The plants left and right of the side knives symbolize cereal rows.

	Ihinger Hof	Kleinhohenheim
	Row Width 150 mm	Row Width 125 mm
Commercial blade width 200 mm	Adjusted blade width 90 mm	Adjusted blade width 70 mm

2.2.3.3 Data Collection

At both research locations, weed plants per m^2 were counted for each plot three days prior to hoeing and the herbicide application. Three days after hoeing, weed plants were counted again for the mechanical treatments at IHO and KH. After a waiting period of three weeks, weed plants were also counted for the herbicide treatment at IHO. Weed counts were made with a quadrat of 1/9 m^2, and 5 samples were collected per plot. An above ground biomass cut of 1 m^2 was performed at BBCH 51 in the spring barley and the spring oats. The plant matter was separated into weed and crop plants. The fresh crop weight was measured shortly after taking the biomass cuts (data not shown). The plant material was then placed in a drying chamber for 48 h at 80 °C. After the drying process, the dry masses of the crop and weed plants were weighed and recorded for each plot. In order to assess the grain yield (t ha^{-1}), only the ten center rows of each plot were harvested. The harvest was performed with two plot combine harvesters. The plots at IHO were harvested on 4 August 2017 with a "Zürn" combine harvester. At KH, the harvest of the oat grains took place on 1 August 2017 with a "Wintersteiger" combine harvester.

2.2.3.4 Data Analysis

The data were analyzed with the R Studio software (Version 1.0.136, RStudio Team, Boston, MA, USA). Prior to the analysis, the data was tested for homogeneity of variance and normal distribution of the residues. An analysis of variance (ANOVA) was performed, and the means of every observation were compared with Duncan's multiple range test at $\alpha \leq 0.05$. The model used was the following:

$$Y_{ijk} = \mu + a_i + b_j + (ab)_{ij} + b_k + e_{ijk}$$

where Y_{ijk} is the result (e.g., grain yield) of treatment i at the driving speed j at block k. μ is the general mean, a_i is the yield attributed to treatment i, b_j is the effect of speed j, $(ab)_{ij}$ is the effect of the interaction between treatment i and speed j, while b_k is the block effect of block k, while e_{ijk} is the residual error of that specific plot.

The weed frequency (%) and weed density (weeds·m^{-2}) were calculated in accordance with Nkoa, Owen and Swanton (2015) as:

$$\text{Frequency} = \frac{\text{no. quadrats with species present}}{\text{no. quadrats}} \times 100\ \%$$

$$\text{Density} = \frac{\sum \text{weed plants present per quadrat}}{\text{no. quadrats}} \times \text{quadrat area}$$

The weed control efficacy (WC) was calculated according to Rasmussen (1991) as:

$$\text{WC} = 100\ \% - \frac{\text{ds}}{0.01 \times \text{du}}$$

where ds is the weed density (weeds·m^{-2}) after application of the treatments and du is the weed density (weeds·m^{-2}) in unweeded control plots.

2.2.4 Results

2.2.4.1 Results at Ihinger Hof

Spring barley yields were higher in the treatments with mechanical weeding compared to the untreated control (Figure 2.2.4-1). The highest spring barley grain yield of 9.8 t ha^{-1} within the mechanical treatments was recorded with the NTS at a 6 km h^{-1} driving speed. The NTS application at 4 km h^{-1} also resulted in higher yields than hoeing with GFS or DSK. The average yield of the untreated control plots was 7.4 t ha^{-1}, which is still a relatively high grain yield for spring barley. The herbicide treatment reached an average yield of 10.3 t ha^{-1}. The treatments NTS(4), NTS(6) and the herbicide application showed significant differences compared to the untreated control. The lowest yield of 8.6 t ha^{-1} for the mechanical treatments was obtained with GFS at 6 km h^{-1}. Reducing the tractor driving speed to 4 km h^{-1} resulted in a crop yield of 9.5 t ha^{-1} for the NTS and 8.9 t ha^{-1} for the GFS treatment. However, the data analysis did not show any significant interaction between treatment and speed, because the differences between each driving speed were only minor. Therefore, no significant grain yield differences could be observed between GFS at 4 and 6 km h^{-1} and NTS at 4 and 6 km h^{-1}. From each plot, grain samples were taken and their moisture contents measured. Averaged over all treatments, the dry substance content of the spring barley was 84.6%. Thus, no significant differences were observed between the treatments concerning grain moisture content. The spring barley dry mass was the highest for the NTS(6) and the lowest for the DSK treatment (Figure 2.2.4-1). The untreated control gained average dry mass yields similar to those of NTS(4) and the herbicide application.

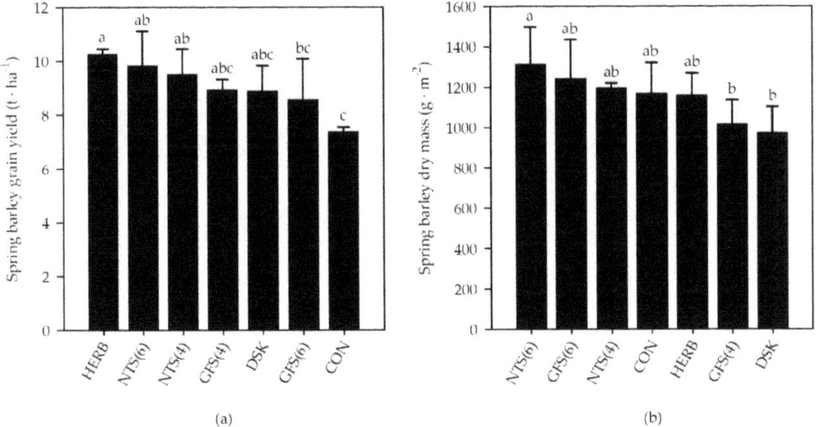

Figure 2.2.4-1: (a) Spring barley grain yield; (b) spring barley dry mass recorded at Ihinger Hof in 2017. Means with the same letter are not significantly different according to Duncan's multiple range test at $\alpha \leq 0.05$. CON = untreated control, HERB = herbicide application, GFS(4) = goosefoot sweeps 4 km h^{-1}, GFS(6) = goosefoot sweeps 6 km h^{-1}, NTS(4) = no-till sweeps 4 km h^{-1}, NTS(6) = no-till sweeps 6 km h^{-1}, DSK = down-cut side knife 4 km h^{-1}.

The weed species composition at IHO showed that no monocot weeds were present during this study (Appendix A Table 2.2.7-2). However, dicot weed species, such as *Chenopodium album* L. and *Convolvulus arvensis* L., were typical for spring cereal cropping systems. All treatments obtained significant weed reduction compared to the untreated control (Figure 2.2.4-2). The untreated control showed a 68.8% increase in weed plants (data not shown). Among the mechanical treatments, no significant differences were observed concerning the weed control efficacy. However, the weed control efficacy of NTS(6) was the lowest (66.8%) and GFS(4) displayed the highest (89.1%) weed control efficacy. The largest decrease of weeds was achieved in the plots treated with the herbicide, with a weed control efficacy of 94.4% on average. The statistical analysis only showed significant differences of the herbicide treatment compared to the NTS(6) treatment. There was no interaction between treatment and speed for GFS and NTS at 4 and 6 km h^{-1}. Despite no interaction between treatment and speed, treatments with GFS(4) showed a weed control efficacy of 89.1%, and hoeing with GFS(6) led to 83.2% weed control efficacy. The no-till sweeps produced weed control efficacies of 76.5% (NTS(4)) and 66.8% (NTS(6)). The weed dry matter was significantly lower for all treatments compared to the untreated control (Figure 2.2.4-2). Inside the untreated control, the weed dry mass was, on average, 25.7 g m^{-2}. All mechanical treatments recorded significantly less dry weed mass. However, none of the mechanical treatments was different from each other. The lowest dry matter data was measured in the GFS(4) and DSK treatments with 2.7 and 2.0 g m^{-2}, respectively. The weed dry matter of the untreated control was more than twelve times higher than the mechanical treatment with DSK. Plots treated with an herbicide produced, on average, 4.8 g m^{-2} of dry weed matter.

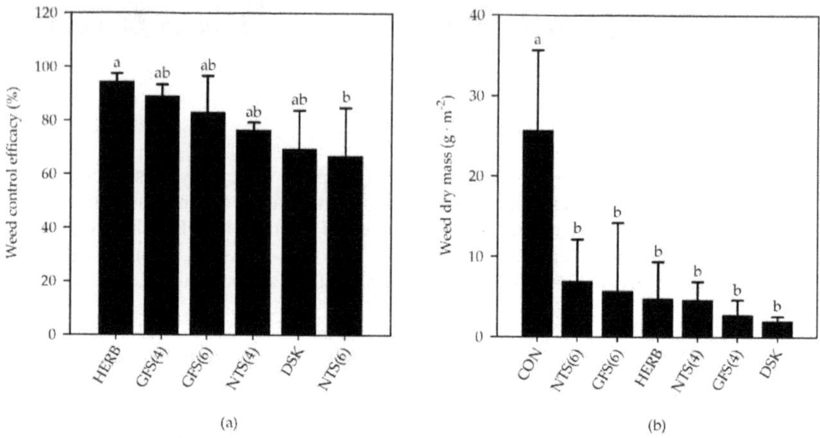

Figure 2.2.4-2: (a) Mean weed control efficacy in spring barley; (b) weed dry mass in spring barley recorded at Ihinger Hof in 2017. Means with the same letter are not significantly different according to Duncan's multiple range test at $\alpha \leq 0.05$. CON = untreated control, HERB = herbicide application, GFS(4) = goosefoot sweeps 4 km h^{-1}, GFS(6) = goosefoot sweeps 6 km h^{-1}, NTS(4) = no-till sweeps 4 km h^{-1}, NTS(6) = no-till sweeps 6 km h^{-1}, DSK = down-cut side knife 4 km h^{-1}.

2.2.4.2 Results at Kleinhohenheim

Compared to the unweeded plots, significant differences in the spring oats grain yield were measured for all treatments except hoeing with GFS(6). The average grain yield of the untreated control was 7.5 t ha^{-1} (Figure 2.2.4-3). All other treatments had a higher average grain yield. No significant interactions existed between treatment and speed when both driving speeds of 4 and 6 km h^{-1} were tested. The highest yield (10.2 t ha^{-1}) was achieved with NTS(4). At a tractor driving speed of 6 km h^{-1}, the yield for the NTS treatment was lower, with an average of 9 t ha^{-1}. Treatments GFS(4) and DSK showed similar high results of 9.5 and 9.2 t ha^{-1}, respectively. Manual weed removal recorded an average grain yield of

9.7 t ha^{-1}. The proportion of grain dry weight was similar for all treatments, with an average of 88.7%. Similarly, to the results at Ihinger Hof, there were no significant statistical differences among treatments concerning the dry crop biomass of all treatments (Figure 2.2.4-3). The average crop dry weights for all treatments ranged from 2100 g m^{-2} (NTS(6)) to 1927 g m^{-2} (NTS(4)).

The weed species composition at KH was different from IHO, but again, it was noticed that no monocot weed species were present (Appendix A Table 2.2.7-2). All mechanical treatments were able to reduce the weed density to a statistical significant extent (Figure 2.2.4-4). The untreated control showed an increase in weed plants between the measurements of up to 52.7% (data not shown). Manual weeding almost removed all of the weeds with a weed control efficacy of 96.9%. The average weed control efficacies of GFS(4), GFS(6) and NTS(6) was 38.6, 50.9 and 47.7% respectively. With 34.3%, hoeing with no-till sweeps at 4 km h^{-1} recorded the lowest weed control efficacy. Inside the untreated control, an average weed dry mass of 34.4 g m^{-2} was measured (Figure 2.2.4-4). Despite having the highest weed control efficacy at Kleinhohenheim, GFS(6) recorded a higher weed dry mass (17.0 g m^{-2}) than the other hoeing treatments. The lowest weed dry mass was measured in the manually weeded plots. The treatments NTS(4), GFS(4) and DSK had similar low dry masses of 10.7, 9.3 and 8.7 g m^{-2}, respectively.

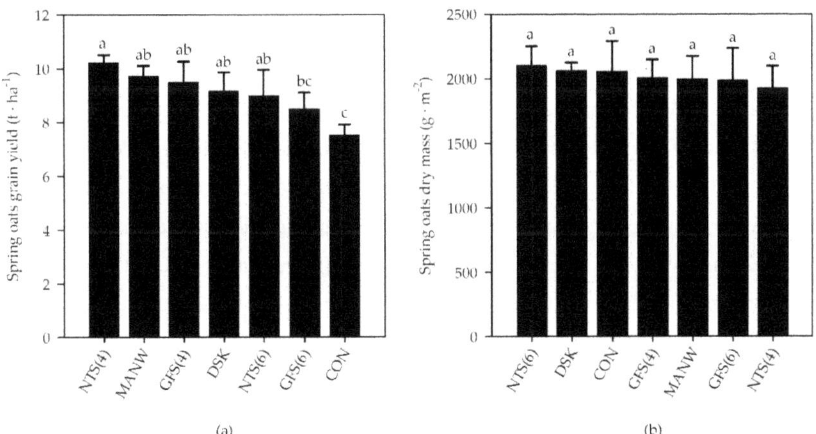

Figure 2.2.4-3: (a) Spring oats grain yield; (b) spring oats dry mass recorded at Kleinhohenheim in 2017. Means with the same letter are not statistically different according to Duncan's multiple range test at $\alpha \leq 0.05$. CON = untreated control, MANW = manual weeding, GFS(4) = goosefoot sweeps 4 km h^{-1}, GFS(6) = goosefoot sweeps 6 km h^{-1}, NTS(4) = no-till sweeps 4 km h^{-1}, NTS(6) = no-till sweeps 6 km h^{-1}, DSK = down-cut side knife 4 km h^{-1}.

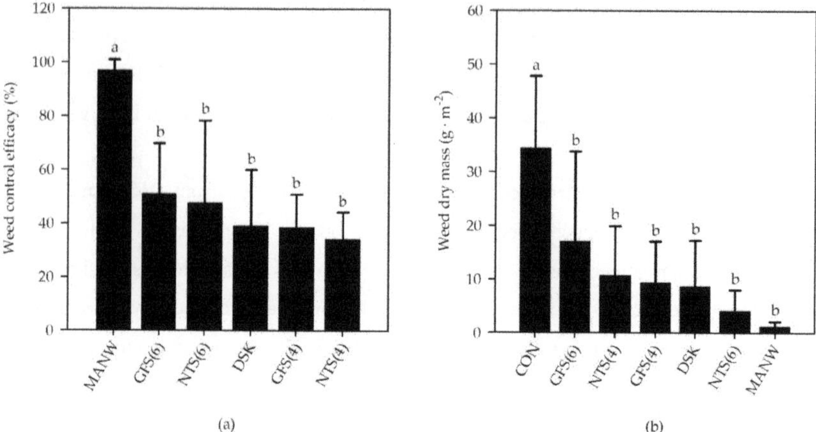

Figure 2.2.4-4: (a) Mean weed control efficacy in spring oats; (b) weed dry mass in spring oats recorded at Kleinhohenheim in 2017. Means with the same letter are not statistically different according to Duncan's multiple range test at $\alpha \leq 0.05$. CON = untreated control, MANW = manual weeding, GFS(4) = goosefoot sweeps 4 km h^{-1}, GFS(6) = goosefoot sweeps 6 km h^{-1}, NTS(4) = no-till sweeps 4 km h^{-1}, NTS(6) = no-till sweeps 6 km h^{-1}, DSK = down-cut side knife 4 km h^{-1}.

2.2.5 Discussion

Hoeing with different sweep types in narrow cereal rows did not decrease grain yields in this experiment. These findings agree with Rasmussen and Svenningsen (1995) who found that hoeing combined with harrowing did not negatively influence cereal crop yield. However, no effect of interaction between driving speed and treatment on the yield results was found in the present study. Melander, Cirujeda and Jørgensen (2003) reported that hoeing with goosefoot sweeps at different driving speeds had no significant effect on winter wheat yields in 240 mm wide row spacing. Paarlberg *et al.* (1998) investigated different cultivation systems in maize, including mechanical weeding. The authors concluded that the hoeing blade style had less distinct effects on yield. Yields were positively influenced by the flat shaped cultivators, similar to the no-till sweeps used in the present experiment. The findings at IHO and KH confirm the tendency that NTS leads to higher grain yields than the other sweep types. Possibly due to limited soil movement into the row, crop plants are less influenced by NTS as they are with GFS at the same speed. This could be especially important for small crop plants, such as cereals. Even though none of the cereal plants in our study were covered by soil, it was observed that goosefoot sweeps moved more soil into the row. The no-till sweeps penetrated the soil well, but due to their flat shape, the soil was neither mixed, nor moved as much, compared to goosefoot sweeps. The same principle applied for hoeing with DSK, due to their flat blade which only cuts through the soil instead of mixing it and the protective vertical blade on one side. The effect of additional soil movement into the intrarow space has to be investigated in further studies. Information about this effect could be obtained by measuring the height of the ridge that is formed by transporting soil into the crop rows (Zhang and Chen, 2017). The yield results in summer barley were exceptionally high for the untreated control, and weeds did not compete with the crop plants, as expected. Much lower yield results for the untreated control were also anticipated at KH. Future experiments could benefit from evenly sowing artificial weeds, such as *Brassica napus* L. or *Sinapis alba* L., as has been done in other studies concerned with mechanical weeding (Rasmussen, 1993; Pullen and Cowell, 1997; Melander, Cirujeda and Jørgensen, 2003; Kolb, Gallandt and Mallory, 2012; Pérez-Ruiz *et al.*, 2015). An increase in weed density could help to accentuate possible effects of different mechanical weeding tools and should be considered. However, sowing of artificial weeds also poses the risk of misjudging the effect of a treatment on a particular weed species. In order to assess the impact of artificial weeds on research results, separate field trials with different densities of artificially sown weeds should be performed. At the moment, it is most important to acknowledge the positive effects of hoeing on grain yields and that narrow row distances are not a limiting factor. Furthermore, even if weed densities are low, mechanical treatments must be carried out to prevent an increase in the weed seed bank and weed dispersal (Rasmussen and Svenningsen, 1995). Furthermore, mechanical cultivation can reduce the use of herbicides and the potential herbicide resistance.

Findings from Melander, Cirujeda and Jørgensen (2003) demonstrated that crop biomass is not significantly affected by intrarow hoeing. Similar results were obtained with the dry weight of the aboveground crop biomass in the current experiment. No significant differences were obtained in the 150 and 125 mm seeded cereals between all treatments compared to the herbicide application and manual weeding. During the experiments, apparent crop damage was not observed. Mechanical weeding did not disrupt plant growth by destroying or uprooting entire barley or oat plants, which could lead to decreased crop biomass. The adjustment of the sweeps to the row widths outside of the experiment in additional trial plots was successful in ensuring optimal hoeing results. The dry crop biomass was not decreased to an extent where hoeing in row spaces of 150 mm and 125 mm would pose an obvious risk for the farmer.

The weed control efficacies of the mechanical treatments ranged from 66.8% to 89.1% at IHO. At KH, hoeing resulted in weed control efficacies of 34.3% to 50.9%. The lower weed control efficacy at KH may be due to the contrasting experimental sites and an overall higher weed count at KH. Goosefoot sweeps resulted in the highest weed control efficacies at IHO and KH. Hoeing with NTS or DSK resulted in lower weed control efficacies, possibly due to less soil being moved into the intrarow space. However, the total weed biomass was reduced effectively at both trial locations by all mechanical treatments. During the experiments, it was observed that the two weed species, *Convolvulus arvensis* L. and *Cirsium arvense* L., were able to regenerate lost plant parts that were cut off during the process of hoeing within 14 days. Therefore, these species pose a great risk concerning effective mechanical weed control, and special attention should be directed towards them throughout the year with suitable farm management strategies. The regenerative ability of *C. arvense* has been addressed by several studies (Hamdoun, 1972; Donald, 1994; Sciegienka, Keren and Menalled, 2011). One approach suggests the cutting of *C. arvense* below the soil surface with goosefoot sweeps (Donald, 1990; Tiley, 2010) to prevent larger root parts from dispersing across the field. During harvest, single *C. arvensis* plants had overgrown the spring barley at IHO and became a harvest issue because the plants got tangled up inside the header of the plot combine harvester. It is therefore strongly suggested that more than one pass with the hoe is performed at the previously suggested hoeing depth of 20 to 30 mm if weed species such as *C. arvensis* and *C. arvense* are present inside the field.

Especially in crops with narrow row spaces, manual steering of a hoe over a long period of time can be tiresome. Crop yields can be decreased due to steering mistakes during the application of mechanical weed control treatments and must be avoided (Home *et al.*, 2002). Therefore, accurate steering of the implements along crop rows is important (Melander and Hartvig, 1997). In the present study it has been shown that hoeing in narrowly spaced cereals is not an issue, and the development of an automatic steering system for these farming conditions seems feasible and is encouraged. The system should have a vision control, composed of a stereo camera to monitor crop rows in front of the hoe. The data provided by the camera can be used to control a hydraulic side shifting frame which follows and adjusts to the crop rows where necessary. A side shifting capability of ±200 mm to the left and right would suffice. For effective weed control, the system should be capable of recognizing cereal rows as early as BBCH 13. The precision of the hoe would be increased, and hoeing could be performed closely to the crop plants. It would also relieve the tractor driver and reduce the risk of crop damage due to steering mistakes of the implement. Furthermore, higher driving speeds can be realized with a camera steered hoe, and more field area can be treated within the same time span (Wilson, 2000; Home *et al.*, 2002; Kunz, Weber and Gerhards, 2015). Additionally, the hoe should be equipped with light emitting diode (LED) spotlights in order to perform weeding during the nighttime. Combining a camera steered system with RTK-GPS (real time kinematics global positioning system) would further increase accuracy and could improve the weeding result. Today, RTK-GPS already has sub-centimeter accuracy (Keicher and Seufert, 2000; Zhang, Wang and Wang, 2002). Together with a camera-guided hoeing frame, the results for the farmer would mean great improvement concerning the area that could be covered by mechanical weeding compared to manual steering of the hoe.

Pullen and Cowell (1997) investigated the interrow performance of different mechanical tools. The authors concluded that a down-cut side knife is the most promising and simplest tool for high-speed hoeing. The present study partially agrees. Pullen and Cowell (1997) performed mechanical treatments at much higher driving speeds (up to 11 km h^{-1}). We agree that DSK-type sweeps can be a good choice for such conditions and may have a high soil mixing effect and a satisfying weed control efficacy, while protecting the crop plants from excess soil coverage. However, while the DSK reached a similar high, but not the highest, yield result as other mechanical treatments, a few disadvantages were observed in the present study. Firstly, two sweeps instead of one are required for hoeing inside the interrow area (Table 2.2.3-3). From a practical perspective, adjusting two implements per interrow area is more time consuming and more complicated than using a single implement, as in, for example, goosefoot or no-till sweeps. Due to the constricted space in the 125 mm row spacing, placement of two side knives between rows was difficult to achieve, because both knives were on one parallelogram and had to fit next to each other in the confined space of the hoeing frame. Side by side mounting was only achieved with difficulty, and a frame with parallelograms placed in offset to each other is recommended for future experiments to avoid space issues when working with sweeps such as side knives in narrow row widths. The second, yet seemingly obvious disadvantage, is that side knives do not carry any soil into the row due to the vertical knife acting as protection for the crop plants (Bowman & Outreach, 2002). Thus, small weed plants which may have been covered by soil with goosefoot or no-still sweeps inside the crop row are spared and can continue competing with crop plants for space, nutrients and water. Finger weeders, used in maize or sugar beets, are not an option for narrow cereal rows because the neighboring crop rows get tangled up in the finger weeders. A third problem that was encountered were crop residues, which the DSK did not always cut through. This led to clogging of soil and plant matter at the 90° bend of the vertical and horizontal knife. Cleaning the knives between each hoe pass was time-consuming, because the second human operator had to dismount from the hoe. Additionally, the shape of the DSK was not optimal for hoeing in grass-like crops, such as cereals. It was observed that single oat and barley leaves could be cut off at the leaf tip due to the down-cut shape of the knives. Even though this occurred only in a few cases and the biomass cuts did not show a reduction of dry mass for the DSK treatment, this could be an issue. Especially in young growth stages, this may lead to negative effects on plant development and yield. Furthermore, mechanical damage to leaves and other plant parts enables pathogens to enter the plants easily and must be avoided.

2.2.6 Conclusions

Hoeing in narrowly spaced cereals has been successfully performed. Further investigations have to show if hoeing with side knives remains a feasible option compared to other sweep types, such as goosefoot and no-till sweeps. Unfortunately, only limited numbers of studies exist which have investigated the effects of different hoeing blade shapes. Future experiments will examine the necessary speed and sweep shape and size (width and length) to achieve optimal yield results and a satisfactory weed control efficacy. A tendency for no-till sweeps to lead to higher average yields than goosefoot sweeps and down-cut side knives was observed. However, goosefoot sweeps generally measured higher weed control efficacies than other mechanical treatments. If there is an effect of driving speed, it has to be investigated by applying a range of speed intensities. Field trials are going to be continued with the presented sweep types at different tractor driving speeds ranging from “slow” (2 km h^{-1}) to “fast” (10 km h^{-1}) in future studies. The effect of multiple hoe passes at different speeds on grain yield, crop biomass and weed control efficacy in narrowly spaced cereals has to be clarified further. The current results suggest that performing consecutive hoe passes could be beneficial in order to achieve higher weed control efficacy, especially for perennial weeds such as *Convolvulus arvensis* L. and *Cirsium arvense* L. Combining the adjusted tools with a camera-based vision system for row recognition and a hydraulically controlled side-shifting frame would facilitate the hoeing process further. A camera-based system might also allow for a smaller safety distance of the sweeps towards the crop rows and would thereby increase the mechanically treated area of the field.

Acknowledgments: The authors would like to thank Sebastian Bökle and Aline Huber for their support during the field experiments at Ihinger Hof. We are also grateful for the assistance of K.U.L.T. Kress Umweltschonende Landtechnik GmbH, Vaihingen an der Enz, Germany and their provision of technical equipment. This project was funded by the German Federal Ministry of Food and Agriculture.

Author Contributions: All authors contributed extensively to this work. Jannis Machleb and Benjamin L. Kollenda organized the execution of the experimental setup, analyzed the data, evaluated the results and drafted the manuscript. Gerassimos G. Peteinatos helped in the data analysis and participated in the draft of the manuscript and its revisions. Roland Gerhards helped in designing the experiment, the drafting and revision of the paper.

Conflicts of Interest: The authors declare no conflict of interest.

2.2.7 Appendix A

Table 2.2.7-1: Overview of the treatments at Ihinger Hof and Kleinhohenheim in 2017.

Treatment	Treatment Description	Driving Speed	Ihinger Hof	Kleinhohenheim
Untreated control	No weed plants removed		x	x
Herbicide treatment	49.5 g a.i. ha^{-1} pinoxaden + 5.5 g a.i. ha^{-1} florasulam (Axial komplett, Syngenta Agro GmbH, Ketsch,Germany), applied with plot sprayer	2.5 km h^{-1}	x	
Manual weeding	Manual weeding			
Goosefoot sweep		4 km h^{-1} and 6 km h^{-1}	x	x
No-till sweep		4 km h^{-1} and 6 km h^{-1}	x	x
Down-cut side knife		4 km h^{-1}	x	x

Naming of the hoeing implements has been done according to (Mohler, 2001; Vincent, Panneton and Fleurat, 2001; Bowman & Outreach, 2002) and personal correspondence with the manufacturer of the implements (K.U.L.T., Vaihingen a. d. Enz). Checks (x) indicate the application of the respective treatment.

Table 2.2.7-2: Weed composition at the two research locations Ihinger Hof and Kleinhohenheim in 2017.

Location	Weed Species	Frequency (%)	Mean Density (Species m^{-2})
Ihinger Hof	*Chenopodium album* L.	78.4	19.1
	Convolvulus arvensis L.	56.3	11.8
	Polygonum convolvulus L.	37.8	5.3
	Thlaspi arvense L.	31.6	7.1
	Others[1]	44.2	4.7
Kleinhohenheim	*Thlaspi arvense* L.	92.9	26.5
	Galium aparine L.	78.6	18.3
	Lamium purpureum L.	61.9	9.9
	Matricaria inodora L.	60.7	17.4
	Sinapis arvensis L.	52.4	11.1
	Polygonum convolvulus L.	41.7	5.9
	Cirsium arvense L.	28.6	6.4
	Others[2]	60.7	7.9

[1]*Cirsium arvense* L., *Veronica persica* L., and *Stellaria media* L.
[2]*Capsella bursa-pastoris* L., *Chenopodium album* L., *Persicaria maculosa* L., *Stellaria media* L. and *Sonchus arvensis* L.

2.2.8 References

BÀRBERI, P. (2002) 'Weed management in organic agriculture: Are we addressing the right issues?', *Weed Research*. Blackwell Science Ltd, pp. 177–193.

BOSTRÖM, U., ANDERSON, L. E. AND WALLENHAMMAR, A. C. (2012) 'Seed distance in relation to row distance: Effect on grain yield and weed biomass in organically grown winter wheat, spring wheat and spring oats', *Field Crops Research*. Elsevier, **134**, pp. 144–152.

BOWMAN, G. AND OUTREACH, S. (2002) *Steel in the field: a farmer's guide to weed management tools, Sustainable Agriculture Network handbook series (USA).*

DONALD, W. (1990) 'Management and control of canada thistle (cirsium arvense)', *Weed science society of America*, **5**, pp. 193–250.

DONALD, W. (1994) 'The biology of Canada thistle (Cirsium arvense)', *Reviews of Weed Science*, pp. 77–101.

FAHAD, S., HUSSAIN, S., CHAUHAN, B. S., SAUD, S., WU, C., HASSAN, S., TANVEER, M., JAN, A. AND HUANG, J. (2015) 'Weed growth and crop yield loss in wheat as influenced by row spacing and weed emergence times', *Crop Protection*. Elsevier, **71**, pp. 101–108.

GALLANDT, E. R., WEINER, J., GALLANDT, E. R. AND WEINER, J. (2015) 'Crop-Weed Competition', in *eLS*. Chichester, UK: John Wiley & Sons, Ltd, pp. 1–9.

HAMDOUN, A. M. (1972) 'Regenerative capacity of root fragments of Cirsium arvense (L.) Scop', *Weed Research*. Blackwell Publishing Ltd, **12**(2), pp. 128–136.

HOME, M. C. W., TILLETT, N. D., HAGUE, T. AND GODWIN, R. J. (2002) 'An experimental study of lateral positional accuracy achieved during inter-row cultivation', in *Proceedings of the 5th EWRS Workshop on Physical and Cultural Weed Control. Scuola Superiore Sant'Anna di studi universitari e di perfezionamento, Pisa, Italy. 11-13 March 2002*, pp. 101–110.

KEICHER, R. AND SEUFERT, H. (2000) 'Automatic guidance for agricultural vehicles in Europe', *Computers and Electronics in Agriculture*, **25**(1–2), pp. 169–194.

KOLB, L. N., GALLANDT, E. R. AND MALLORY, E. B. (2012) 'Impact of Spring Wheat Planting Density, Row Spacing, and Mechanical Weed Control on Yield, Grain Protein, and Economic Return in Maine', *Weed Science*. Weed Science Society of America 810 East 10th Street, Lawrence, KS 66044-8897, **60**(02), pp. 244–253.

KOUWENHOVEN, J. K. AND TERPSTRA, R. (1979) 'Sorting action of tines and tine-like tools in the field', *Journal of Agricultural Engineering Research*. Academic Press, **24**(1), pp. 95–113.

KUNZ, C., WEBER, J. F., PETEINATOS, G. G., SÖKEFELD, M. AND GERHARDS, R. (2017) 'Camera steered mechanical weed control in sugar beet, maize and soybean', *Precision Agriculture*.

KUNZ, C., WEBER, J. AND GERHARDS, R. (2015) 'Benefits of Precision Farming Technologies for Mechanical Weed Control in Soybean and Sugar Beet—Comparison of Precision Hoeing with Conventional Mechanical Weed Control', *Agronomy*. Multidisciplinary Digital Publishing Institute, **5**(2), pp. 130–142.

LANCASHIRE, P. D., BLEIHOLDER, H., BOOM, T. VAN DEN, LANGELÜDDEKE, P., STAUSS, R., WEBER, E. AND WITZENBERGER, A. (1991) 'A uniform decimal code for growth stages of crops and weeds', *Annals of Applied Biology*. John Wiley & Sons, Ltd (10.1111), **119**(3), pp. 561–601.

MATTSSON, B., NYLANDER, C. AND ASCARD, J. (1990) 'Comparison of seven inter-row weeders', *3rd International Conference IFOAM, Non-chemical Weed Control*, pp. 91–107.

MELANDER, B., CIRUJEDA, A. AND JØRGENSEN, M. H. (2003) 'Effects of inter-row hoeing and fertilizer placement on weed growth and yield of winter wheat', *Weed Research*. Blackwell Science Ltd, **43**(6), pp. 428–438.

MELANDER, B. AND HARTVIG, P. (1997) 'Yield responses of weed-free seeded onions [Allium cepa (L.)] to hoeing close to the row', *Crop Protection*, **16**(7), pp. 687–691.

MOHLER, C. L. (2001) 'Enhancing the competitive ability of crops', in *Ecological Management of Agricultural Weeds*, pp. 269–321.

NKOA, R., OWEN, M. D. K. AND SWANTON, C. J. (2015) 'Weed Abundance, Distribution, Diversity, and Community Analyses', *Weed Science*, **63**(SP1), pp. 64–90.

PAARLBERG, K. R., HANNA, H. M., ERBACH, D. C. AND HARTZLER, R. G. (1998) 'Cultivator design for interrow weed control in no-till corn', *Applied Engineering in Agriculture*. American Society of Agricultural and Biological Engineers, **14**(4), pp. 353–361.

PÉREZ-RUIZ, M., GONZALEZ-DE-SANTOS, P., RIBEIRO, A., FERNANDEZ-QUINTANILLA, C., PERUZZI, A., VIERI, M., TOMIC, S. AND AGÜERA, J. (2015) 'Highlights and preliminary results for autonomous crop protection', *Computers and Electronics in Agriculture*. Elsevier, pp. 150–161.

PULLEN, D. W. M. AND COWELL, P. A. (1997) 'An Evaluation of the Performance of Mechanical Weeding Mechanisms for use in High Speed Inter-Row Weeding of Arable Crops', *Journal of Agricultural Engineering Research*. Academic Press, **67**(1), pp. 27–34.

RASMUSSEN, J. (1991) 'A model for prediction of yield response in weed harrowing', *Weed Research*, **31**(6), pp. 401–408.

RASMUSSEN, J. (1992) 'Testing harrows for mechanical control of annual weeds in agricultural crops', *Weed Research*. Blackwell Publishing Ltd, **32**(4), pp. 267–274.

RASMUSSEN, J. (1993) 'Yield response models for mechanical weed control by harrowing at early crop growth stages in peas (Pisum sativum L.)', *Weed Research*, **33**(3), pp. 231–240.

RASMUSSEN, J. AND SVENNINGSEN, T. (1995) 'Selective weed harrowing in cereals', *Biological Agriculture and Horticulture*. Taylor & Francis Group, **12**(1), pp. 29–46.

SCIEGIENKA, J., KEREN, E. AND MENALLED, F. (2011) 'Impact of root fragment dimension, weight, burial depth, and water regime on Cirsium arvense emergence and growth', *Canadian Journal of Plant Science*, **91**(6), pp. 1027–1036.

TERPSTRA, R. AND KOUWENHOVEN, J. K. (1981) 'Inter-row and intra-row weed control with a hoe-ridger', *Journal of Agricultural Engineering Research*. Academic Press, **26**(2), pp. 127–134.

TILEY, G. E. D. (2010) 'Biological Flora of the British Isles: Cirsium arvense (L.) Scop.', *Journal of Ecology*. Blackwell Publishing Ltd, **98**(4), pp. 938–983.

TILLETT, N. D. AND HAGUE, T. (1999) 'Computer-Vision-based Hoe Guidance for Cereals — an Initial Trial', *Journal of Agricultural Engineering Research*. Academic Press, **74**(3), pp. 225–236.

TILLETT, N. D. AND HAGUE, T. (2006) 'Increasing Work Rate in Vision Guided Precision Banded Operations', *Biosystems Engineering*. Academic Press, **94**(4), pp. 487–494.

TILLETT, N., HAGUE, T., GARFORD, P. AND WATTS, P. (2003) 'Development of a commercial vision guided inter-rowhoe: achievements, problems and future directions', in *Proceedings of the 4 th European Conference on Precision Agriculture ECPA, Berlin*, pp. 671–676.

VINCENT, C., PANNETON, B. AND FLEURAT, F. (2001) *Physical Control Methods in Plant Protection*. Springer Science & Business Media.

WELSH, J. P., TILLETT, N. D., HOME, M. AND KING, J. A. (2002) 'A review of knowledge: Inter-row hoeing and its associated agronomy in organic cereal and pulse crops', *Research Policy*, **15**, pp. 1–21.

WILSON, J. N. (2000) 'Guidance of agricultural vehicles—a historical perspective', *Computers and electronics in agriculture*, **25**(1), pp. 3–9.

WILTSHIRE, J. J. J., TILLETT, N. D. AND HAGUE, T. (2003) 'Agronomic evaluation of precise mechanical hoeing and chemical weed control in sugar beet', *Weed Research*. Blackwell Science Ltd, **43**(4), pp. 236–244.

ZHANG, N., WANG, M. AND WANG, N. (2002) 'Precision agriculture—a worldwide overview', *Computers and Electronics in Agriculture*, **36**, pp. 113–132.

ZHANG, X. AND CHEN, Y. (2017) 'Soil disturbance and cutting forces of four different sweeps for mechanical weeding', *Soil and Tillage Research*, **168**, pp. 167–175.

2.3 Sensor-based intrarow mechanical weed control in sugar beets with motorized finger weeders

J MACHLEB*, G G PETEINATOS†, M SÖKEFELD† & R GERHARDS†

*Department of Weed Science, Institute of Phytomedicine, University of Hohenheim, Stuttgart
†Department of Weed Science, Institute of Phytomedicine, University of Hohenheim, Stuttgart

Running head: Sensor-based mechanical weed control

*Correspondence: Jannis Machleb, University of Hohenheim, Otto-Sander-Str. 5, 70599 Stuttgart, Tel: (+49) 711 459 22394; E-mail: jmachleb@uni-hohenheim.de

Submitted to: *Weed Research*.

2.3.1 Summary

Increased attention has been dedicated to the development of agricultural robots for autonomous weeding in the past years. This also requires the development of suitable mechanical weeding tools. Therefore, we devised a new weeding tool for agricultural robots to perform intrarow mechanical weed control in sugar beets and other wide row crops. A conventional finger weeder was modified and equipped with an electric motor. This allowed the rotational movement of the finger weeders independent of the forward travel speed of the tool carrier. The new tool was tested in combination with a bi-spectral camera in a two-year field trial. The camera was used to identify crop plants in the intrarow area. A controller regulated between two different rotational speeds of the motorized finger weeder. At the location of a sugar beet plant, the rotational speed was equal to the driving speed of the tractor. Between two sugar beets, the rotational speed was either increased by 40% or decreased by 40%. The intrarow weed control efficacy of this new system ranged between 87% to 91% in 2017 and 91% to 94% in 2018. Sugar beet yields were not adversely affected by the mechanical treatments compared to the conventional herbicide application.

The motorized finger weeders present an effective system for selective intrarow mechanical weeding. Certainly, mechanical weeding involves the risk of high weed infestations if the treatments are not applied properly and timely regardless if sensor technology is used or not. However, due to the increase of herbicide resistances and the continuing bans on herbicides mechanical weeding strategies must be investigated further. The mechanical weeding system of the present study can contribute to the reduction of herbicide use in sugar beets and other wide row crops.

Keywords: robotic weeding, row crops, sensor-assisted weeding, weed management, hoeing, mechanical weeding

2.3.2 Introduction

Herbicides have replaced the majority of weed control methods since their introduction in the middle of the 20th century (Hall *et al.*, 2006; Glaeser, 2011). However, due to a combination of legal constraints, public demand, and environmental concerns, mechanical weed control has re-emerged as an effective alternative to the application of synthetic herbicides (Buhler, 1996). New mechanical weeding tools and precise steering techniques evolved rapidly during the past years (Merfield, 2019).

Especially by combining already existing implements for mechanical weeding with a variety of sensor systems increased their utilization in arable crops. For example, camera-guided hoes are readily available on the markets from different manufacturers. Their guidance concept is based on a camera that tracks the crop rows and sends a signal to a hydraulic cylinder which shifts the hoe left or right in order to stay aligned with the crop rows (Slaughter *et al.*, 1999). The benefits of such systems include: reduced driver fatigue, the possibility to work closer to the crop which results in an increase of the area that is treated, and the potential to increase the working speed (Wilson, 2000). Field experiments with camera-steered hoes in wide row crops have shown that the weed densities could be reduced by up to 87% (sugar beets) and 89% (soybean) (Kunz, Weber and Gerhards, 2015).

Depending on their field of application, mechanical weed control methods can be divided into intrarow and inter-row treatments. The intrarow area is the small strip of the crop rows themselves whereas the inter-row area is the space between two adjacent crop rows (Pérez-Ruiz *et al.*, 2012). Today, post-emergent inter-row weeding can be performed reliably in wide row crops without excessive crop damage, however, the removal of intrarow weeds is still a challenging task (Cloutier *et al.*, 2007; Tillett *et al.*, 2008). Therefore, a variety of different tools have been developed to physically deal with intrarow weeds. This includes finger weeders, torsion weeders and weeding brushes (Rasmussen and Svenningsen, 1995; Kouwenhoven, 1997).

In the course of automation, mechanical weed control in arable crops has also become of interest in combination with agricultural robots. Bosch Deepfield Robotics developed the BoniRob which eliminates individual weed plants with a stamp tool (Langsenkamp *et al.*, 2014). Robots have also been tested successfully to eradicate *Rumex obtusifolius* L. (broad leaved dock) in pastures (van Evert *et al.*, 2011). The AgBotII combines mechanical and chemical weed control implements (Bawden *et al.*, 2017) and a combination of thermal and mechanical tools has been evaluated with robots of the RHEA fleet in Spain (Pérez-Ruiz *et al.*, 2015). However, there is still room for improvement in the design and set-up of the tools for mechanical weed control with robots.

This study examined a novel approach to mechanical weeding in sugar beets (*Beta vulgaris* subsp. *vulgaris*, Altissima Group). Sugar beets are not very competitive with weeds and they must be kept weed free until row-closure (Cioni and Maines, 2010). In Germany, two to three post-emergent herbicide applications are usually necessary for sufficient weed control in sugar beets. Mechanical weeding is not practiced often due to the high risk of weed infestations if the treatments are not performed properly and timely. Intrarow weeds in particular are difficult to remove because sugar beets are sensitive plants and do not tolerate physical stress well. Therefore, we have developed an imaging system which determines the position of sugar beets to perform intrarow mechanical weed control with a motorized finger weeder. The motorized finger weeder is based on a conventional finger weeder which was equipped with an electric motor. This enabled a rapid variation of the rotational movement of the finger weeder independent of the tractor driving speed. The system was integrated into a conventional camera-steered inter-row hoeing system for row crops. It was tested in a two-year field experiment with sugar beets and compared to conventional mechanical and chemical weeding methods. It was hypothesized that a) the motorized finger weeder increased the weed control efficacy compared to the conventional finger weeder and that b) sugar beet yields are not negatively impacted by the motorized finger weeders compared to the conventional finger weeders. Furthermore, the combination of inter-row and intrarow mechanical weed control provided equal weed control efficacies to herbicide applications across the whole plots.

2.3.3 Materials and Methods

2.3.3.1 Overview and Experimental Site

Field trials for intrarow mechanical weed control were carried out in sugar beets in the years 2017 and 2018. The experiment was located at the trial site Ihinger Hof (Renningen, south-west Germany) and consisted of 8 different treatments and 4 replications. The Ihinger Hof trial site is 475 m above sea level with an average long-term rainfall of 738 mm. Mechanical weed control methods with motorized finger weeders (MFW) were compared to an untreated control, a conventional herbicide application and ground-driven conventional finger weeders (CFW). A detailed description of the treatments can be found in Table 2.3.3-1. The experimental design was a randomized complete block design in both trial years.

Table 2.3.3-1: Description of the treatments at Ihinger Hof in 2017 and 2018.

Treatment	Treatment acronyms	Description
Untreated control	UC	No weed control
Herbicide 2017	H2017	Herbicide spraying, see Table 2.3.3-2
Herbicide 2018	H2018	Herbicide spraying, see Table 2.3.3-2
Motorized finger weeder *Fast speed*	MFW(140)	3 x Motorized finger weeding with 40% higher rotational speed between two sugar beets than at the location of a sugar beet
Herbicide and motorized finger weeder *Fast speed*	H+MFW(140)	1 x Herbicide spraying 2 x Motorized finger weeding with 40% higher rotational speed between two sugar beets than at the location of a sugar beet
Motorized finger weeder *Slow speed*	MFW(60)	3 x Motorized finger weeding with 40% lower rotational speed between two sugar beets than at the location of a sugar beet
Herbicide and motorized finger weeder *Slow speed*	H+MFW(60)	1 x Herbicide spraying 2 x Motorized finger weeding with 40% lower rotational speed between two sugar beets than at the location of a sugar beet
Conventional finger weeder	CFW	3 x Conventional finger weeding
Herbicide and conventional finger weeder	H+CFW	1 x Herbicide spraying 2 x Conventional finger weeding

The tractor driving speed was 1 km h^{-1} for the MFW treatments and 6 km h^{-1} for the CFW treatments. The nominal rotational speed (100%) of the MFW at the location of a sugar beet corresponded to a linear speed of 1 km h^{-1}.

In both trial years, sugar beets cv. Hannibal were sown 3 cm deep with a 3 m wide seeder (Solitair 8, Lemken, Alpen, Germany) at a row distance of 0.5 m. The seeding density was 107 000 seeds ha^{-1} which results in a distance of 18 to 22 cm between two beet plants. The plot size was 12 m x 3 m (length x width). Thus, each plot comprised six sugar beet rows and the sowing width (3 m) matched the hoeing width (3 m).

2.3.3.2 Herbicides and application details

Table 2.3.3-2 lists the herbicides that were used in the experiments in 2017 and 2018. The three chemical-mechanical treatments H+MFW(140), H+MFW(60) and H+CFW were sprayed once with the herbicides listed for BBCH 10 (Lancashire *et al.*, 1991) of the sugar beets. Due to the high weed density in 2017, three herbicide applications (BBCH 10, 14 and 18) were applied in the H2017 treatment. In 2018, only two herbicide applications (BBCH 10 and 14) were necessary to control the weeds.

The herbicide application was performed with a battery-powered plot sprayer (Schachtner, Ludwigsburg, Germany). The spray boom of the sprayer was 3 m wide and equipped with flat spray nozzles (TWIN flat spray air-injector compact nozzles IDKT 120-05, Lechler, Metzingen, Germany). The herbicides were applied with 300 L water ha^{-1} with a spray pressure of 200 kPa and 50 cm above ground level at a driving speed of 5 km h^{-1}.

Table 2.3.3-2: Herbicide type and application time (BBCH of the sugar beets) at Ihinger Hof in 2017 and 2018.

BBCH 2017	BBCH 2018	Common name	Product name	Formulation	Concentration	Application rate	Supplier
10, 14, 18	10, 14	Desmedipham	Betanal® maxxPro®	OD	47 g a.i. L^{-1}	70.5 g ha^{-1}	Bayer CropScience
10, 14, 18	10, 14	Phenmedipham	Betanal® maxxPro®	OD	60 g a.i. L^{-1}	90 g ha^{-1}	Bayer CropScience
10, 14, 18	10, 14	Ethofumesate	Betanal® maxxPro®	OD	75 g a.i. L^{-1}	112.5 g ha^{-1}	Bayer CropScience
10, 14, 18	10, 14	Lenacil	Betanal® maxxPro®	OD	27 g a.i. L^{-1}	40.5 g ha^{-1}	Bayer CropScience
10, 14, 18	10, 14	Metamitron	Goltix® Titan®	SC	525 g a.i. L^{-1}	630 g ha^{-1}	ADAMA Germany
10, 14, 18	10, 14	Quinmerac	Goltix® Titan®	SC	40 g a.i. L^{-1}	48 g ha^{-1}	ADAMA Germany
14	14	Fluazifop-P-butyl	Fusilade Max®	EC	107 g a.i. L^{-1}	107 g ha^{-1}	Nufarm Germany
14, 18	14	Clopyralid	Lontrel™ 720	SG	720 g a.i. kg^{-1}	165 g ha^{-1}	Dow AgroSciences

OD - oil dispersion; SG - water-soluble granules; SC - suspension concentrate; EC - emulsifiable concentrate;
Spray volume: 300 L water ha^{-1}
Each herbicide application received 0.5 L ha^{-1} of the additive Oleo FC (94% paraffin oils and 6% emulsifiers, ADAMA Germany)

2.3.3.3 General set-up of the hoe

The set-up of the mechanical weed control implement (Figure 2.3.3-1) was based on a 3 m wide hoeing frame (Argus, K.U.L.T., Vaihingen a. d. Enz, Germany) in combination with a Garford Robocrop Side Shift System (Garford Farm Machinery Ltd., Peterborough, England). The hoeing system had two cameras. The first camera was the Garford camera for general row alignment of the hoe to perform interrow hoeing. The second camera was our bi-spectral camera which was situated above a sugar beet row and responsible for the intrarow treatments with the motorized finger weeders.

Hoeing between the sugar beet rows was performed with 20 cm wide goosefoot sweeps mounted on a parallelogram. Since each plot comprised 6 sugar beet rows, seven parallelograms were required to treat all interrow spaces of one plot. The safety distance towards the sugar beet rows was set to be 5 cm. For intrarow weeding, one pair of conventional finger weeders was used per sugar beet row. Prior to each application, the CFW and MFW were adjusted outside of the experiment to ensure an optimal weeding result.

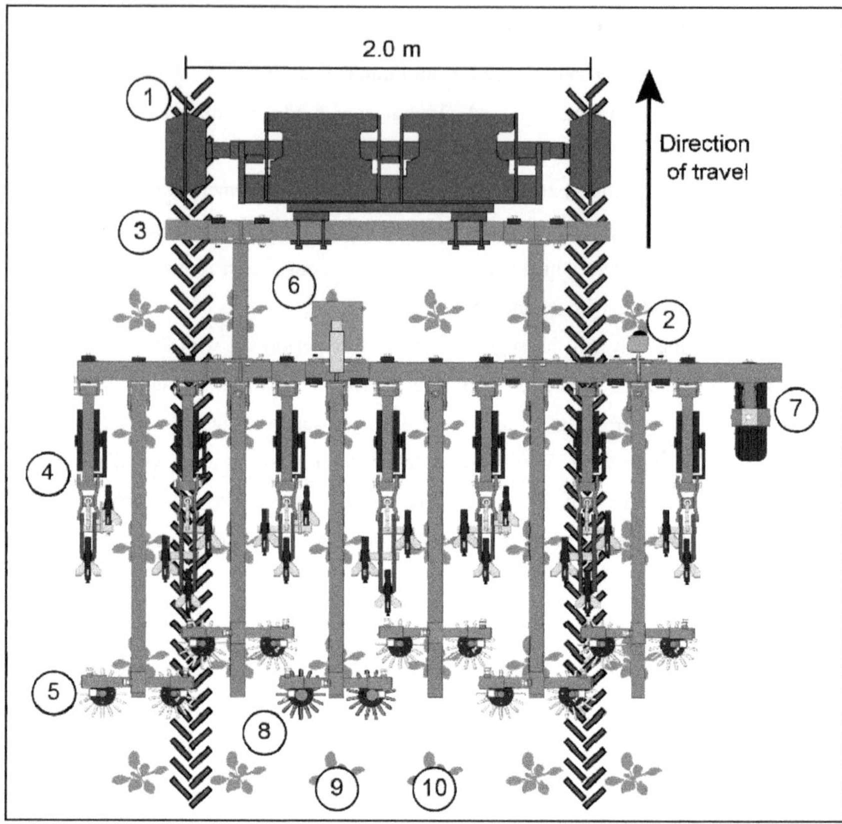

Figure 2.3.3-1: The set-up of the hoe used in the experiments at Ihinger Hof in 2017 and 2018. The image displays one plot width (six sugar beet rows). 1 = Garford Robocrop, 2 = Garford Robocrop camera for interrow weeding, 3 = Argus hoe frame, 4 = parallelograms with goosefoot sweeps, 5 = conventional finger weeders, 6 = bi-spectral camera for intrarow weeding (colour blue), 7 = odometry wheel, 8 = motorized finger weeders (colour blue), 9 & 10 = the two centre sugar beet rows of each plot that were treated with the MFW and used for harvesting. Design ele-ments by K.U.L.T. Germany.

2.3.3.4 Set-up of the motorized finger weeders

The idea behind the bi-spectral camera in combination with the MFW was to switch between two rotational speeds during the treatment. The distance between two sugar beets was supposed to be treated with a different rotational speed than the area at the location of a sugar beet. Two different treatment intensities (slow and fast) of the strip between two sugar beet plants were tested in this study (Figure 2.3.3-2). On the one hand, this served to test whether the sensor system could reliably switch between different speeds and, on the other hand, whether there was a difference in weed control performance between the two rotational speeds.

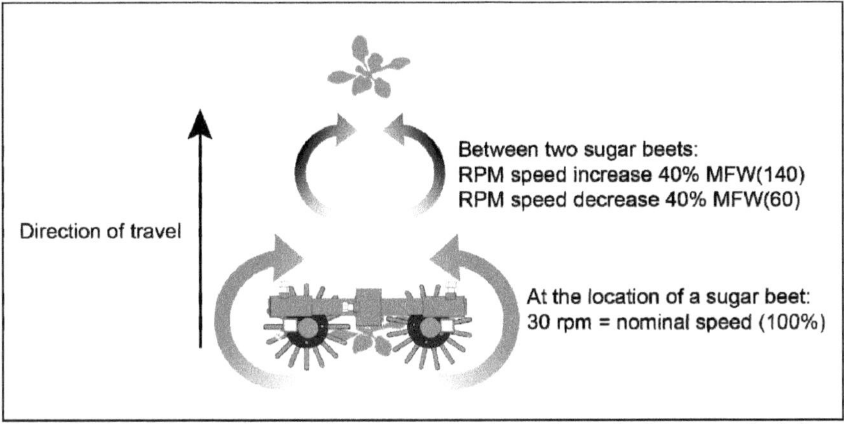

Figure 2.3.3-2: Diagram showing the rotation direction and speed of the motorized finger weeders.

For the construction of the motorized finger weeders, one pair of CFW was modified (Figure 2.3.3-3). The metal spines which normally propel the CFW as the tractor drives forward were dismounted. Then an electric motor (Bühler Motor GmbH, Germany) was mounted on the plastic disc of each finger weeder. The finger weeders had an external diameter of 24 cm and an internal diameter of 12 cm. Since this prototype had never been tested in a field experiment, only one pair of MFW was assembled and tested for its weed control performance. The motors used were the PM12 which was a voltage regulated DC gear motor. They were equipped with a two-stage gear giving a 19:1 speed reduction. Therefore, each motor provided around 6 kg m^2 s^{-2} maximum rated torque. The maximum speed on the soil was 45 rpm when the motor was supplied with 24V DC. Since the motors were propelling the two finger weeders, they were placed in such a way that they were approaching in the left and right of the row. Therefore, the motors were configured to rotate clockwise and counterclockwise respectively. A dual-channel motor controller, namely the Robotec MDC2460 (Robotec Inc, Scottsdale, USA) was responsible for controlling the speed and the rotating motion of the two motors.

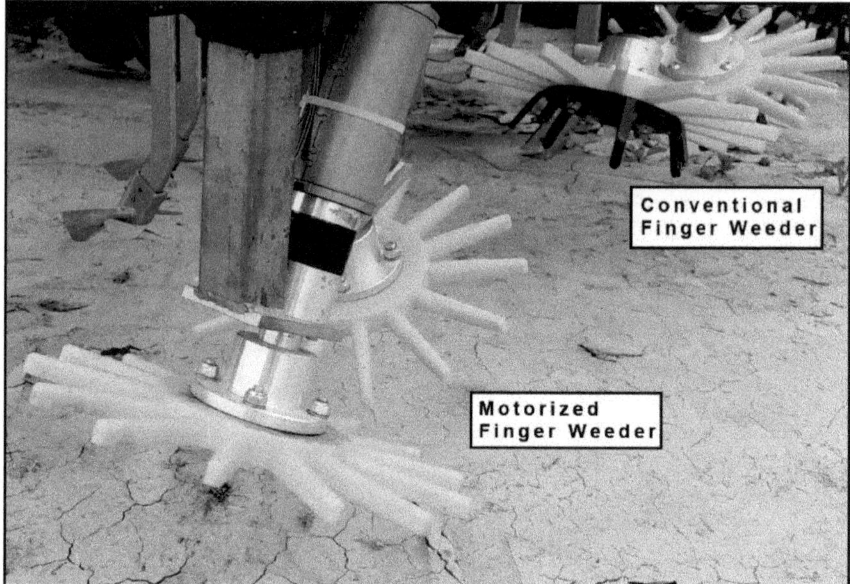

Figure 2.3.3-3: The motorized finger weeders (front) and conventional finger weeders (back).

The typical speed of the motors at the location of a sugar beet plant was around 30 rpm when placed on the ground. Therefore, the middle of the finger weeder (φ 180 mm) would achieve a linear speed of 1km h^{-1} which was concurrent with the speed that the tractor was moving and is referred to as the nominal speed (100%) from hereon. By rotating with the same speed as the tractor was moving forward, the MFW worked unhindered without causing any damage to the sugar beet plants. The treatments MFW(140) and H+MFW(140) had the MFW rotate with a speed 40% faster than the driving speed of the tractor between two sugar beet plants. In this scenario, it was expected that the damage and the uprooting of the weeds was higher due to the more active and enforced collision of the MFW. A higher soil disturbance from the faster rotation was also expected.

In the treatments MFW(60) and H+MFW(60), the finger weeders rotated with a speed 40% slower between two sugar beet plants than that of the nominal speed applied at the location of a sugar beet. Thus, the rotational movement of the finger weeder was slower than the forward travel speed of the tractor (1 km h^{-1}). This forced the finger weeders to perform a little sliding motion in the direction of travel. If this caused changes in the weed control efficacy compared to MFW(140) and H+MFW(140) was part of this study.

2.3.3.5 Implementation of the mechanical treatments

One pair of CFW was replaced by the MFW for plots which received treatment with MFW. Only the two centre sugar beet rows of each plot were treated with the MFW. Hoeing was performed either two or three times, depending on the treatment. Table 2.3.3-3 lists the mechanical treatments and their corresponding time of application as the BBCH growth stage of the sugar beets.

Table 2.3.3-3: Time of application of the mechanical treatments at Ihinger Hof according to the number of developed leaves and the BBCH stage (Lancashire *et al.*, 1991).

Treatment	Treatment acronyms	Sugar beet growth stage	BBCH growth stage
Motorized finger weeder *Fast speed*	MFW(140)	3 to 4 leaves 4 to 5 leaves 6 to 8 leaves	13 - 14 14 - 15 16 - 18
Herbicide and motorized finger weeder *Fast speed*	H+MFW(140)	4 to 5 leaves 6 to 8 leaves	14 - 15 16 - 18
Motorized finger weeder *Slow speed*	MFW(60)	3 to 4 leaves 4 to 5 leaves 6 to 8 leaves	13 - 14 14 - 15 16 - 18
Herbicide and motorized finger weeder *Slow speed*	H+MFW(60)	4 to 5 leaves 6 to 8 leaves	14 - 15 16 - 18
Conventional finger weeder	CFW	3 to 4 leaves 4 to 5 leaves 6 to 8 leaves	13 - 14 14 - 15 16 - 18
Herbicide and conventional finger weeder	H+CFW	4 to 5 leaves 6 to 8 leaves	14 - 15 16 - 18

Since only one pair of MFW existed, two passes with the tractor were necessary to treat two sugar beet rows in each plot. The sowing pattern of the sugar beets was advantageous because both centre rows were offset left and right from the actual centre of the plot. Therefore, it was possible to perform hoeing with the bi-spectral camera and the MFW without having to change their position on the hoeing frame. The tractor simply had to turn after the first pass with the hoe and drive through the plot again from the other direction. However, two consecutive passes with the goosefoot sweeps had to be avoided. Otherwise, the plots receiving MFW-treatments would also receive double the amount of interrow passes with the goosefoot sweeps which would have distorted the weed control results. Therefore, the parallelograms with the goosefoot sweeps were lifted and locked into their floating position prior to driving through the plot a second time with the MFW. The treatments with CFW did not require two passes since enough conventional finger weeders were available for this weeding method. The harvest of the CFW-plots was also restricted to the two centre rows in order to achieve comparable yield results.

2.3.3.6 Description of the sensor system

In this experiment, a red-infrared camera was used for individual crop plant recognition. The sensor was a camera-based machine vision system with autonomous illumination in both red and infrared, making the sensor usable even at night or low illumination conditions. A modified version of the software IMPASS was used for the image identification and plant species classification.

The bi-spectral camera takes two pixel-congruent images in the red and near-infrared spectrum. Differential images (infrared-red) were calculated out of the two images. Thresholding was automatically performed, creating a binary image where the plant material was attributed as 1 (white) and the rest of the objects were attributed as background with a value of 0 (black). Due to the reflection characteristics of the different materials, the difference image does not contain any disturbances like stones, straw or other organic matter. In addition, a strong contrast between plants and background could be achieved (Sökefeld *et al.*, 2007). For the weed classification, a data-based image analysis system was used. In the database, parameters (shape features and morphological features) of the different weed species and sugar beets are stored. The features of the objects found in the images were compared with the features of the model plants stored in the database. They were classified in the appropriate class by a minimum distance classifier (Weis and

Gerhards, 2007). Each complete white object in the image was separated and based on the shape recognition attributes it was classified as sugar beet or weed. Then, each image was separated in strips of around 40 mm for the direction of movement and for each strip a decision was made if it contained a sugar beet plant or if it contained a weed.

Figure 2.3.3-1 and Figure 2.3.3-4 describe the set-up of the sensor-system that was used in this study. The bi-spectral camera Figure 2.3.3-5 was mounted onto the hoe, facing downwards on one of the two centre sugar beet rows. The principle of the setup was to separate the intrarow region into 40 mm strips. Since the space between the sugar beets was typically around 180 mm, this space had at least 3 sometimes even 4 strips without the presence of a sugar beet and then a strip that contained a sugar beet. At the strip with the sugar beet, the nominal speed was applied. At the rest of the strips, the speed was adjusted according to the treatment (faster or slower respectively). A wheel was mounted on the hoe, in order to measure the distance travelled by the hoe. If no movement was monitored from the wheel, the motors were halted to reduce the wear on the motors and the finger weeders, and to avoid possible crop damage. A Raspberry Pi Model B coordinated the different tasks. Based on the input from the wheel encoder the Raspberry Pi triggered the sensor and activated IMPASS to process the image and make the necessary decisions. These decisions were returned to the Raspberry Pi. In order to avoid untimely reactions, the image analysis was performed in another computer (ThinkPad Lenovo P50, Intel Core i7-6700HQ at 2.60 GHz, 32 GB of RAM & a Quadro M100M PCIe graphics card). An odometry measurement was necessary to measure the travel distance of the hoe in order to synchronize the movement of the MFW with the position of the crop plants. Based on the odometry measurements and the image analysis, the Raspberry-Pi sent a signal to the Roboteq microcontroller to regulate the movement speed of the MFW.

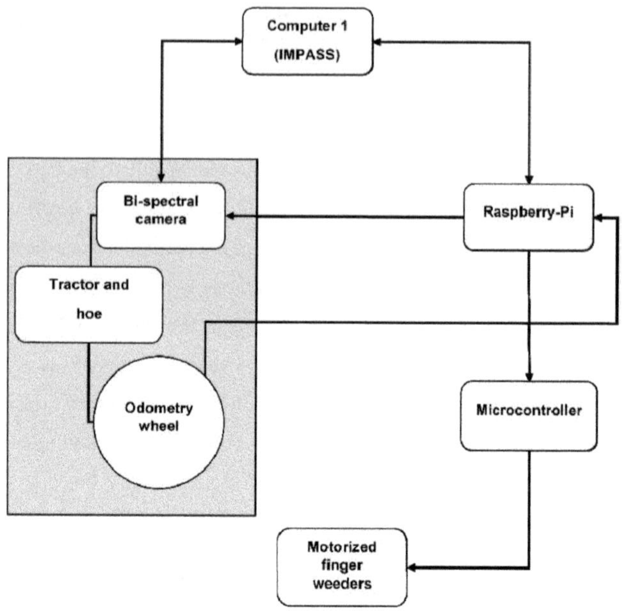

Figure 2.3.3-4: Flowchart describing the sensor system.

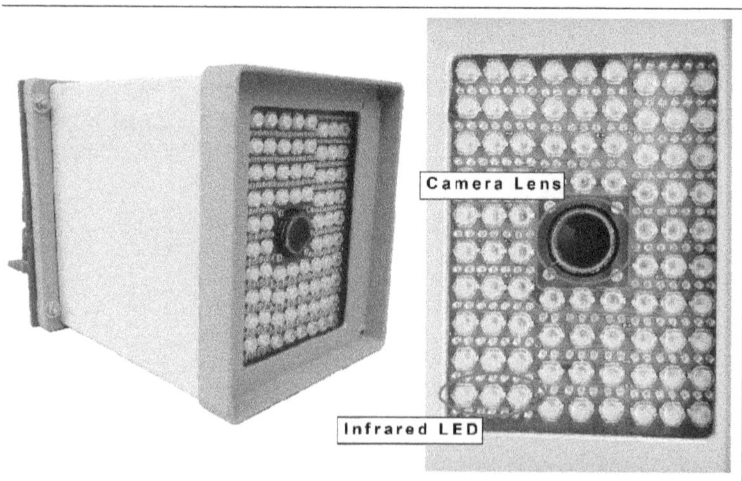

Figure 2.3.3-5: The bi-spectral camera for intrarow weeding: the lens is surrounded by the infrared LEDs.

2.3.3.7 Data acquisition

The weed density (plants m-2) was measured by using a 0.5 m x 0.5 m frame. The frame was divided into an intrarow and inter-row section to differentiate between both areas (Figure 2.3.3-6). The frame was placed randomly at three locations inside the centre of each plot three days after the final application of the mechanical treatments. The sugar beet harvest took place by the end of September in both trial years. A plot harvester uprooted and collected the sugar beets. All 12 m of the two centre sugar beet rows of each plot were harvested.

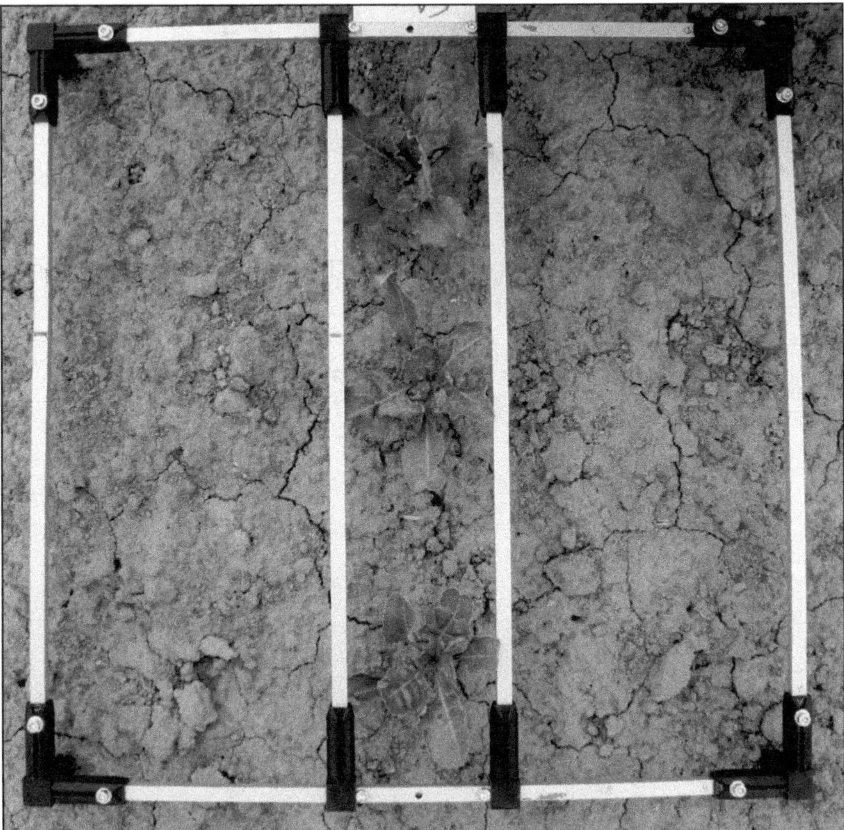

Figure 2.3.3-6: The frame (0.5 x 0.5 m) that was used for the measurements of the weed density in 2017 and 2018.

2.3.3.8 Data analysis

The data were analysed with RStudio (R Version 3.3.1) as a randomized complete block design. Prior to the analysis of variance (ANOVA), the data were checked for homogeneity of variance and normal distribution of the residuals. The means of the observations were compared with the Tukey HSD-Test at $\alpha \leq 0.05$. The model used was the following:

$$Y_{ijk} = \mu + a_i + b_j + (ab)_{ij} + b_k + e_{ijk}$$

where Y_{ijk} is the result (e.g. sugar beet yield) of the treatment i at the driving speed j at block k. μ is the general mean, a_i is the yield attributed to treatment i, b_j is the effect of speed j, $(ab)_{ij}$ is the effect of the interaction between treatment i and speed j, while b_k is the block effect of block k, while e_{ijk} is the residual error of that specific plot.

The weed density (plants m^{-2}) in each plot was calculated according to Nkoa, Owen and Swanton (2015) as:

$$\text{Density} = \frac{\sum \text{weed plants present per quadrat}}{\text{no. quadrats}} \div \text{quadrat area}$$

The weed control efficacy (WCE in %) is a measure for the effectiveness of a treatment to reduce the weed density. For each treatment and plot, the WCE was calculated according to Rasmussen (2006) as:

$$\text{WCE (\%)} = 100 - \frac{d_t}{0.01 \times d_u}$$

where d_t is the weed density (plants m^{-2}) after application of the treatments and d_u is the weed density (plants m^{-2}) in the untreated control plots.

2.3.4 Results

2.3.4.1 Weed Density and Weed Control Efficacy

There was a high variation in the weed density between 2017 and 2018. In 2017, the weed densities in the untreated control plots reached on average 588 plants m^{-2}, whereas the mean weed density in 2018 was 38 plants m^{-2}. The divergence in weed density was due to a wet spring season in 2017 with 654 mm of total annual rainfall, whereas only 526 mm of precipitation occurred in 2018. Additionally, the trial field in 2017 provided a larger weed seedbank than expected. Thus, it was decided to analyse the data from 2017 and 2018 separately. However, the weed species found in both fields were typical for the local sugar beet production. The most frequent weed species were *Chenopodium album* L. (common lambsquarters), *Matricaria inodora* L. (scentless false mayweed), *Cirsium arvense* L. (creeping thistle), *Galium aparine* L. (cleavers) and *Polygonum convolvulus* L. (black bindweed).

The results in Table 2.3.4-1 show that the herbicide treatments were the most effective method to reduce the weed density in both trial years. The highest weed densities were recorded in the untreated control plots, ranging from 588 plants m^{-2} in 2017 to 43 plants m^{-2} in 2018. Inter-row hoeing with the goosefoot sweeps eliminated most of the weed plants growing between the sugar beet rows and no significant differences were found between any of the treatments. In 2017, weed densities of 1.7 and 1.3 plants m^{-2} were recorded for the H+MFW(140) and the H+MFW(60) treatments respectively. However, those findings were not significantly different compared to the pure herbicide treatment. Slightly higher weed densities of 5.3 plants m^{-2} were found in the H+CFW treatment, but the results were not significantly different compared to the combined methods H+MFW(140) and H+MFW(60). The mechanical treatments without an additional herbicide application (MFW(140), MFW(60) and CFW) showed significantly higher total weed densities than treatments that received a single herbicide application. In 2018, the differences between mechanical treatments in combination with an herbicide and without an herbicide were not as prominent as in 2017. The average intrarow weed densities after application of the treatments ranged from 1.7 to 6.8 plants m^{-2}.

Table 2.3.4-1: The results obtained for the mean weed density (plants m^{-2}) of the intrarow and inter-row area measured three days after the final application of each treatment in 2017 and 2018. Additionally, the results for the intrarow weed control efficacy (%) are shown. Means with different letters within the same column indicate significant differences between the treatments according to the Tukey HSD-Test at $\alpha \leq 0.05$. WCE = weed control efficacy.

	2017				**2018**			
	Weed density (weeds m^{-2})				**Weed density (weeds m^{-2})**			
Treatment*	**intrarow**	**inter-row**	**sum**	**Intrarow WCE (%)**	**intrarow**	**inter-row**	**sum**	**Intrarow WCE (%)**
UC	177.0^{a}	411.0^{a}	588.0^{a}	-	29.5^{a}	13.5^{a}	43.0^{a}	-
H2017/18	0.0^{c}	0.0^{b}	0.0^{e}	100^{a}	0.0^{d}	0.0^{b}	0.0^{c}	100^{a}
MFW(140)	15.0^{bc}	3.3^{b}	18.3^{c}	92^{bc}	2.7^{bcd}	0.5^{b}	3.2^{bc}	91^{ab}
H+MFW(140)	1.7^{c}	0.0^{b}	1.7^{de}	99^{ab}	1.7^{cd}	0.3^{b}	2.0^{bc}	94^{a}
MFW(60)	19^{bc}	0.3^{b}	19.3^{c}	89^{c}	2.3^{bcd}	0.6^{b}	2.9^{bc}	92^{ab}
H+MFW(60)	1.3^{c}	0.0^{b}	1.3^{de}	99^{ab}	1.7^{cd}	0.1^{b}	1.8^{bc}	94^{a}
CFW	37^{b}	6.0^{b}	43^{b}	79^{d}	6.8^{b}	0.4^{b}	7.2^{b}	77^{c}
H+CFW	4.3^{c}	1.0^{b}	5.3^{d}	98^{ab}	6.5^{bc}	0.3^{b}	6.8^{b}	78^{bc}

*see Table *2.3.3-1 for a detailed treatment description*

The highest intrarow weed density was found in the untreated control in both trial years. However, the difference between the average weed density of the untreated control in 2017 and 2018 was 147.5 plants m^{-2}. No significant differences were obtained within the mechanical weed control treatments. Though, conventional finger weeders recorded an average intrarow weed density of 37 plants m^{-2} whereas MFW(140) and MFW(60) led to lower weed densities of 15 and 19 plants m^{-2}, respectively in 2017.

Table 2.3.4-1 also shows the mean intrarow weed control efficacy (WCE) for the years 2017 and 2018 of each treatment. The lowest weed control efficacy was recorded for the conventional finger weeders in 2017 (79%) and also in 2018 with a weed control efficacy of 77 to 78% for H+CFW and CFW. The highest weed control efficacy was achieved with the herbicide treatments H2017 and H2018. In 2017, the combination of mechanical and chemical treatments showed similarly high WCE levels as the herbicide application.

2.3.4.2 Sugar beet yield

The lowest sugar beet yield (Figure 2.3.4-1) was found in the untreated control (2017) with an average yield of 6.9 t ha^{-1}. Higher yields of 57.2 t ha^{-1} for the untreated control were recorded in 2018. The four treatments H2017, H+MFW(140), H+MFW(60) and H+CFW achieved similarly high sugar beet yields between 75 and 77.9 t ha^{-1}. Among the purely mechanical weed control methods, MFW(140) and MFW(60) yielded between 48.5 and 52.4 t ha^{-1} and CFW recorded 36.2 t ha^{-1}. No statistically significant differences were found between the average sugar beet yields of the mechanical and chemical-mechanical treatments in 2018. Table 2.3.4-2 provides an overview of the number of sugar beets per hectare that were present prior to the treatments (crop emergence) and at harvest. As with all mechanical treatments some crop plant losses were observed for all mechanical treatments. However, the overall yields did not differ from the conventional herbicide treatment in 2017 and 2018. The number of sugar beets for the untreated control in 2017 decreased by 40 000 until autumn because weed competition was so high that only half of the sugar beets were in a harvestable condition. The data also shows that less sugar beets emerged in 2018 than in 2017. This was probably due to a lack of rainfall in spring.

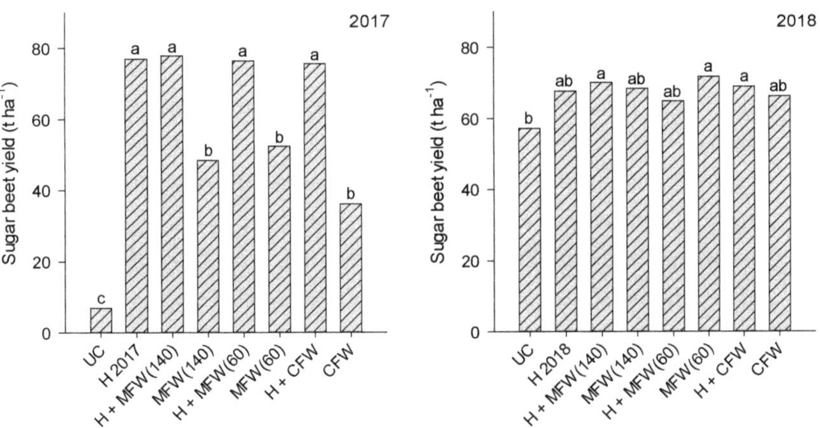

Figure 2.3.4-1: The bars represent the mean sugar beet yield (t ha^{-1}) recorded for each treatment in 2017 and 2018. Different letters above a bar and within the same graph (2017 or 2018) indicate significant differences between the treatments according to the Tukey HSD-Test at $\alpha \leq 0.05$.

Table 2.3.4-2: The number of sugar beets per hectare before the application of the treatments (crop emergence) and at the harvest date in 2017 and 2018.

	Sugar beets ha^{-1} before treatments		Sugar beets ha^{-1} at harvest	
Treatment*	**2017**	**2018**	**2017**	**2018**
UC	87842	80051	48125	79791
Herbicide	87532	80446	87509	80416
MFW(140)	86546	79258	86455	78333
H+MFW(140)	87622	80042	87312	79554
MFW(60)	87331	80427	86959	78103
H+MFW(60)	88429	79802	87514	79166
CFW	87596	81113	86980	80401
H+CFW	86580	78728	86233	78354

**see* Table *2.3.3-1 for a detailed treatment description*

2.3.5 Discussion

Inter-row hoeing has been proven to be an effective post-emergent weed control method in wide row crops (van Zuydam, Sonneveld and Naber, 1995; Tillett *et al.*, 2002; Wiltshire, Tillett and Hague, 2003). High weed control efficacy results can be obtained if the mechanical treatments are applied at the right stage of the weed development and under good weather conditions (De Buck *et al.*, 1999). This could be confirmed in the present study. Furthermore, using the Garford steering-system provided a high guidance accuracy across all mechanical treatments. Post-emergent hoeing with goosefoot sweeps was very successful and recorded average inter-row weed control efficacies between 94% and 98% (Table 2.3.4-1). Mechanical intrarow weeding with the motorized and the conventional finger weeders also demonstrated weed control efficacies nearly as high as the conventional herbicide applications in both years. Only in case of a very high weed pressure, as observed in 2017, the combination of low-dose herbicide application with post-emergent finger weeding resulted in a higher weed control efficacy than mechanical weeding alone.

Like all post-emergent mechanical weeding operations, finger-weeding in sugar beets requires advanced crop development relative to weed development. Selectivity of finger weeding is low, if weeds emerge earlier than the sugar beets (Hatcher and Melander, 2003). Advanced crop development relative to weed growth can be achieved in different ways. Applying a pre-emergent herbicide in low doses is the most common way to suppress early emerging weeds in sugar beets. False seedbed preparation and shallow pre-emergent harrowing also prevent weeds emerging before the crop (Dierauer and Stöppler-Zimmer, 1994). Later emerging weeds can then be removed with mechanical weeding tools at a higher intensity. In years with low weed densities, herbicide applications could be substituted partially or even completely by mechanical treatments as the results from 2018 demonstrate. Weeding with the MFW with and without spraying produced similarly high weed control efficacies as the conventional herbicide application in 2018. Thus, two out of three herbicide sprayings were saved in 2018. Band spraying was not performed in this study but it presents another option to reduce the herbicide input even further (Kouwenhoven, Wevers and Post, 1991) by savings of 50% (Pérez-Ruiz *et al.*, 2013) to 65% (Kunz *et al.*, 2016) when compared to overall spraying.

In 2018, sugar beet yields and white sugar yields (data not shown) were not adversely affected by the mechanical treatments compared to three times spraying. The low yield results of the MFW and the CFW in 2017 reflect the difficulties that come with high weed densities. Sugar beets simply do not tolerate weeds growing close-by due to their low competitiveness for resources (Heisel, Andreasen and Christensen, 2002). Petersen (2004) stated that yield losses in sugar beets can reach 95% if no plant protection measures are performed. This study confirmed these findings because the untreated control recorded a yield decrease of 91% compared to the most successful treatment (herbicide + motorized finger weeder with 140 rpm) in 2017. Since not all intrarow weeds were removed from the sugar beet rows by the motorized finger weeders or the conventional finger weeders, yield losses due to crop-weed competition were inevitable in 2017.

Finger weeders are highly effective and can reduce weed densities by up to 99% (Riemens *et al.*, 2007). Therefore, this basic concept was adapted for the new autonomous weeding tool in the present study. Additionally, it was important to use a rotating tool similar to the cycloid hoe with metal tines used by Åstrand and Baerveldt (2002). However, finger weeders could be the better choice because they are made from rubber plastic. This makes them flexible yet tough and they are not as rigid as steel tools. Other rotating tools like brushes can have trouble penetrating hard and heavy soils, thus, leading to insufficient weed control (Kouwenhoven, 1997). It must be stated that the motorized finger weeders had a slow working speed compared to conventional farming implements. This was due to the technological limitations of the system. However, the concept of the motorized finger weeders is not supposed to be used with a tractor but rather in conjunction with a robot to constantly perform autonomous weeding. Currently, agricultural robots work slowly compared to a tractor. This is an issue because conventional implements for mechanical weeding require a certain amount of speed to effectively cut, uproot or bury weed plants (Machleb *et al.*, 2020). For example, inter-row hoeing is most effective at speeds of 4 to 12 km h^{-1} (Bowman and Outreach, 2002). Therefore, mechanical weeding tools which rotate independently of the tool carrier (robot) are the optimal solution for autonomous robotic weeding because the rotating movement compensates the slower forward travel speed. Hereby, the accurate guidance of the tool is essential to ensure alignment with the crop row to prevent crop damage. This was achieved successfully with the Garford steering-system in the present study.

The motorized finger weeders performed well in the loamy soils at Ihinger Hof. However, no interaction was found between the rotational speed and the weed control efficacy of the motorized finger weeders. Recognition of the sugar beets and adjusting the speed of the MFW was performed reliably by the data provided from the infrared camera and the IMPASS software. An image separation between weed and crop suffices if only mechanical weeding is performed. Our algorithm is based on shape-detection and it was able to differentiate well between weeds and sugar beets. The motorized finger weeders require that the entire crop row is monitored so that the previously defined areas of a sugar beet and the space between two sugar beets can be targeted precisely with the implement. This is similar to the concept of using images (e.g. aerial data) that can be used to compute detailed herbicide application maps for specific weed species to perform site-specific spraying (Mink *et al.*, 2018). However, if more plant groups must be separated instead of just crop and weed plants, other deep learning algorithms are necessary. A promising alternative are convolutional neural networks which use raw pixel values instead of feature descriptors (Dyrmann, Karstoft and Midtiby, 2016).

Even though the motorized finger weeders usually resulted in a higher weed control efficacy than the conventional finger weeders, no significant differences were found among the mechanical treatments concerning the sugar beet yield in 2017 and 2018. Overall, the motorized finger weeders proved to be an effective tool for intrarow mechanical weed control in sugar beets concerning the weed control efficacy. It can be concluded that the motorized finger weeders performed better than the conventional finger weeders and they could be an ideal tool for mechanical weeding with an agricultural robot. A robotic system with motorized finger weeders or other implements can easily compensate for the slow working speed, due to more robots working simultaneously (concept of a swarm of robots). Furthermore, the system is not restricted to sugar beets alone and can be transferred to other crops in further studies. This may include vegetables (e.g. lettuce or cabbage) where the plant spacing of one crop plant to the other is large enough.

Other approaches to robotic weed control are systems which apply small amounts of herbicides to single weed plants by using computer vision and machine learning to differentiate between weed and crop plants. Blue River Technology (California, USA) developed such a system. It could be improved further if chemical and mechanical treatments were combined. The system could decide whether a weed can be removed with the mechanical tool or if it must be sprayed. For example, weeds growing very close to the crop can be difficult to remove mechanically without causing crop damage and should be sprayed.

An improvement of the motorized finger weeders would be to use even smaller finger weeders with a diameter of less than 20 cm. This would increase their precision further because the space between two sugar beet plants is only about 18 to 22 cm wide. Higher and lower rotational speeds should also be tested in the future to determine if even better weed control results can be achieved. In general, robotic mechanical weeding is seen as a promising alternative to conventional herbicide applications and should be part of an integrated weed management concept. However, one prerequisite is that the robot can work continuously to compensate for its slow working speed. Additional benefits include that a robot is lightweight and causes less soil compaction than a tractor.

2.3.6 Acknowledgements

We would like to thank the technicians at Ihinger Hof for their support during the trial season 2017 and 2018. Furthermore, we thank Dr. Peter Risser and Dr. Johann Maier from Südzucker AG Germany for the technical support and funding of this project.

2.3.7 Conflict of Interest

The authors declare no conflict of interest.

2.3.8 References

ÅSTRAND, B. AND BAERVELDT, A. J. (2002) An agricultural mobile robot with vision-based perception for mechanical weed control. *Autonomous Robots* **13**(1)pp. 21–35.

BAWDEN, O., KULK, J., RUSSELL, R., MCCOOL, C., ENGLISH, A., DAYOUB, F., LEHNERT, C. AND PEREZ, T. (2017) Robot for weed species plant-specific management. *Journal of Field Robotics* **34**(6)pp. 1179–1199.

BOWMAN, G. AND OUTREACH, S. (2002) *Steel in the Field. A Farmer's Guide to Weed Management Tools. Network*. Available at: http://agris.fao.org/agris-search/search.do?recordID=US19970148386 (Accessed: 6 February 2018).

DE BUCK, A. J., SCHOORLEMMER, H. B., WOSSINK, G. A. A. AND JANSSENS, S. R. M. (1999) Risks of post-emergence Weed control strategies in Sugar beet: Development and application of a bio-economic model. *Agricultural Systems*. Elsevier **59**(3)pp. 283–299.

BUHLER, D. D. (1996) Development of alternative weed management strategies. in *Journal of Production Agriculture*. American Society of Agronomy, Crop Science Society of America, Soil Science Society of America pp. 501–505.

CIONI, F. AND MAINES, G. (2010) Weed Control in Sugarbeet. *Sugar Tech* pp. 243–255.

CLOUTIER, D. C., VAN DER WEIDE, R. Y., PERUZZI, A. AND LEBLANC, M. L. (2007) 8 Mechanical Weed Management. *Non-chemical Weed management* p. 111.

DIERAUER, H.-U. AND STÖPPLER-ZIMMER, H. (1994) *Unkrautregulierung ohne Chemie*. Ulmer.

DYRMANN, M., KARSTOFT, H. AND MIDTIBY, H. S. (2016) Plant species classification using deep convolutional neural network. *Biosystems Engineering* **151**pp. 72–80.

VAN EVERT, F. K., SAMSOM, J., POLDER, G., VIJN, M., DOOREN, H. J. VAN, LAMAKER, A., VAN DER HEIJDEN, G. W. A. M., KEMPENAAR, C., VAN DER ZALM, T. AND LOTZ, L. A. P. (2011) A robot to detect and control broad-leaved dock (Rumex obtusifolius L.) in grassland. *Journal of Field Robotics* **28**(2)pp. 264–277.

GLAESER, B. (2011) *The Green Revolution revisited : critique and alternatives*. Routledge.

HALL, J. C., VAN EERD, L. L., MILLER, S. D., OWEN, M. D. K., PRATHER, T. S., SHANER, D. L., SINGH, M., VAUGHN, K. C. AND WELLER, S. C. (2006) Future Research Directions for Weed Science 1. *Weed Technology* **14**(3)pp. 647–658.

HATCHER, P. E. AND MELANDER, B. (2003) Combining physical, cultural and biological methods: Prospects for integrated non-chemical weed management strategies. *Weed Research* pp. 303–322.

HEISEL, T., ANDREASEN, C. AND CHRISTENSEN, S. (2002) Sugarbeet yield response to competition from Sinapis arvensis or Lolium perenne growing at three different distances from the beet and removed at various times during early growth. *Weed Research* **42**(5)pp. 406–413.

KOUWENHOVEN, J. K. (1997) Intra-row mechanical weed control - Possibilities and problems. *Soil and Tillage Research* **41**(1–2)pp. 87–104.

KOUWENHOVEN, J. K., WEVERS, J. D. A. AND POST, B. J. (1991) Possibilities of mechanical post-emergence weed control in sugar beet. *Soil and Tillage Research*. Elsevier **21**(1–2)pp. 85–95.

KUNZ, C., STURM, D. J., PETEINATOS, G. G. AND GERHARDS, R. (2016) Weed Suppression of Living Mulch in Sugar Beets. *Gesunde Pflanzen* **68**(3)pp. 145–154.

KUNZ, C., WEBER, J. AND GERHARDS, R. (2015) Benefits of Precision Farming Technologies for Mechanical Weed Control in Soybean and Sugar Beet—Comparison of Precision Hoeing with Conventional Mechanical Weed Control. *Agronomy* **5**(2)pp. 130–142.

LANCASHIRE, P. D., BLEIHOLDER, H., BOOM, T. VAN DEN, LANGELÜDDEKE, P., STAUSS, R., WEBER, E. AND WITZENBERGER, A. (1991) A uniform decimal code for growth stages of crops and weeds. *Annals of Applied Biology* **119**(3)pp. 561–601.

LANGSENKAMP, F., SELLMANN, F., KOHLBRECHER, M., KIELHORN, A., STROTHMANN, W., MICHAELS, A., RUCKELSHAUSEN, A. AND TRAUTZ, D. (2014) Tube Stamp for mechanical intra-row individual Plant Weed Control. *Agricultural Engineering International: the CIGR Ejournal* pp. 1–11.

MACHLEB, J., PETEINATOS, G. G., KOLLENDA, B. L., ANDÚJAR, D. AND GERHARDS, R. (2020) Sensor-based mechanical weed control: Present state and prospects. *Computers and Electronics in Agriculture*. Elsevier **176**p. 105638.

MERFIELD, C. N. (2019) Integrated Weed Management in Organic Farming. in *Organic Farming*. Woodhead Publishing pp. 117–180.

MINK, R., DUTTA, A., PETEINATOS, G. G., SÖKEFELD, M., ENGELS, J. J., HAHN, M. AND GERHARDS, R. (2018) Multi-temporal site-specific weed control of Cirsium arvense (L.) scop. and Rumex crispus L. in maize and sugar beet using unmanned aerial vehicle based mapping. *Agriculture (Switzerland)* **8**(5).

NKOA, R., OWEN, M. D. K. AND SWANTON, C. J. (2015) Weed Abundance, Distribution, Diversity, and Community Analyses. *Weed Science* **63**(SP1)pp. 64–90.

PÉREZ-RUIZ, M., CARBALLIDO, J., AGÜERA, J. AND RODRÍGUEZ-LIZANA, A. (2013) Development and evaluation of a combined cultivator and band sprayer with a row-centering RTK-GPS guidance system. *Sensors (Switzerland)* **13**(3)pp. 3313–3330.

PÉREZ-RUIZ, M., GONZALEZ-DE-SANTOS, P., RIBEIRO, A., FERNÁNDEZ-QUINTANILLA, C., PERUZZI, A., VIERI, M. AND AGÜERA, J. (2015) Highlights and preliminary results for autonomous crop protection. *Computers and Electronics in Agriculture* **110**pp. 150–161.

PÉREZ-RUIZ, M., SLAUGHTER, D. C., GLIEVER, C. J. AND UPADHYAYA, S. K. (2012) Automatic GPS-based intra-row weed knife control system for transplanted row crops. *Computers and Electronics in Agriculture*. Elsevier pp. 41–49.

PETERSEN, J. (2004) A Review on Weed Control in Sugarbeet. in *Weed Biology and Management*. Dordrecht: Springer Netherlands pp. 467–483.

RASMUSSEN, J. (2006) A model for prediction of yield response in weed harrowing. *Weed Research* **31**(6)pp. 401–408.

RASMUSSEN, J. AND SVENNINGSEN, T. (1995) Selective weed harrowing in cereals. *Biological Agriculture and Horticulture* **12**(1)pp. 29–46.

RIEMENS, M. M., VAN DER WEIDE, R. Y., BLEEKER, P. O. AND LOTZ, L. A. P. (2007) Effect of stale seedbed preparations and subsequent weed control in lettuce (cv. Iceboll) on weed densities. *Weed Research*. John Wiley & Sons, Ltd (10.1111) **47**(2)pp. 149–156.

SLAUGHTER, D., CHEN, P., AGRICULTURE, R. C.-P. AND 1999, U. (1999) Vision Guided Precision Cultivation. *Precision Agriculture* **1**(2)pp. 199–216.

SÖKEFELD, M., GERHARDS, R., OEBEL, H. AND THERBURG, R. D. (2007) Image acquisition for weed detection and identification by digital image analysis. in *Precision Agriculture 2007 - Papers Presented at the 6th European Conference on Precision Agriculture, ECPA 2007* pp. 523–528.

TILLETT, N., HAGUE, T., AGRICULTURE, S. M.-C. AND ELECTRONICS IN AND 2002, U. (2002) Inter-row vision guidance for mechanical weed control in sugar beet. *Computers and Electronics in Agriculture* **33**(3)pp. 163–177.

TILLETT, N., HAGUE, T., GRUNDY, A., ENGINEERING, A. D.-B. AND 2008, U. (2008) Mechanical within-row weed control for transplanted crops using computer vision. *Biosystems Engineering* **99**(2)pp. 171–178.

WEIS, M. AND GERHARDS, R. (2007) Feature extraction for the identification of weed species in digital images for the purpose of site-specific weed control. *Precision agriculture* **7**pp. 537–545.

WILSON, J. N. (2000) Guidance of agricultural vehicles - A historical perspective. *Computers and Electronics in Agriculture* **25**(1–2)pp. 3–9.

WILTSHIRE, J. J. J., TILLETT, N. D. AND HAGUE, T. (2003) Agronomic evaluation of precise mechanical hoeing and chemical weed control in sugar beet. *Weed Research* **43**(4)pp. 236–244.

VAN ZUYDAM, R. P., SONNEVELD, C. AND NABER, H. (1995) Weed control in sugar beet by precision guided implements. *Crop Protection* **14**(4)pp. 335–340.

3 GENERAL DISCUSSION

This thesis discusses different aspects of modern, sensor-based mechanical weed control and it investigates the implementation of new weed control methods for sensor-based mechanical weeding in the precision agriculture sector. The review paper "Sensor-based mechanical weed control: Present state and prospects" provides a comprehensive overview of sensor applications for mechanical weeding and how agriculture can profit from the use of such systems. Additionally, the implications of utilizing sensors in an agricultural environment are discussed. The second paper "Adjustment of weed hoeing to narrowly spaced cereals" assesses the feasibility of hoeing in narrow row widths which are typical for European conventional cereal farming systems. It analyzes the effects of interrow hoeing based on the weed control efficacy and cereal yield. Lastly the development and testing of a new system for sensor-guided intrarow mechanical weeding in sugar beet is discussed with the paper "Sensor-based intrarow mechanical weed control in sugar beets with motorized finger weeders". The following discussion will summarize the findings of each paper, highlight, and discuss assets and drawbacks of mechanical weeding and provide prospects for future research and developments from which modern agriculture may profit.

Even though the first paper "**Sensor-based mechanical weed control: Present state and prospects**" focuses on modern sensor-technology, it was equally important to provide an agronomic context for the implications that come with mechanical weed control in arable crops. Mechanical weed control is a complex area of agriculture, which is the reason why it has been somewhat neglected during the past decades due to the preferred use of herbicides in many cropping systems. However, starting in the 1980s, a new approach to plant protection evolved with integrated weed management (IWM) practices due to the growing concern over possible negative effects of herbicides on the environment and public health (Swanton and Weise, 1991; Kogan, 1998; Moss, 2010; Özkara, Akyil and Konuk, 2016). This approach includes alternative direct weed control strategies such as mechanical weeding for conventional farming systems to reduce the input of herbicides. However, mechanical weed control poses several implications that must be considered. For example, favorable soil and weather conditions must be awaited. Suitable conditions would be sunny and dry weather, and friable soil because uprooted weed plants need to dry out to prevent their regrowth. Furthermore, the time frame for mechanical weeding is very small because the treatments should be applied at an early weed growth stage. The smaller the weed plants are at the time of the treatments, the higher are the chances of a high weed control level. Harrows, for example, bury and uproot small weed seedlings but their efficacy decreases rapidly as weed plants grow taller and become more anchored in the soil (Kurstjens and Perdok, 2000; Kurstjens and Kropff, 2001). The same observation has been made by Vangessel *et al.* (1998) who state that early mechanical weed control in dry beans (*Phaseolus vulgaris* L.) resulted in the lowest weed populations. Therefore, mechanical weed control usually targets weeds emerging prior to crop establishment and shortly after crop emergence (Hatcher and Melander, 2003).

However, this leads to a predicament because targeting weeds during the early crop growth stages increases the risk of damaging the delicate crop plants. Thus, it is important to consider the tolerance of the crop to withstand the physical forces of the weeding tools and the treatment intensity must be adjusted accordingly (Hatcher and Melander, 2003). This factor must always be taken into account and so far, the review article has shown that mechanical weed control must still rely on the operator to set up the machine and the implement for weeding to avoid unnecessary crop damage. An ideal scenario would be to have a sensor-based mechanical weeding implement, which is able to detect the crop growth stage on its own. It should then adjust the treatment intensity accordingly. Usually, the treatment intensity is not kept static so that it can be increased over time as the crop grows and becomes more tolerant of mechanical stress and soil burial (Rasmussen, 1992). But it is advisable to test a new treatment intensity carefully prior to treating an entire field to achieve acceptable weed control levels without excessive crop damage. A certain amount of crop damage is unproblematic in crops like cereals, (Rasmussen, 1992; Rasmussen and Svenningsen, 1995; Rasmussen, Bibby and Schou, 2008). However, delicate crops like sugar beets may suffer yield losses if treated too intense. Therefore, avoiding crop plant damage during the critical early growth stages of sensitive crops is the most important aspect of mechanical weed control to avoid direct crop losses (Kouwenhoven, 1997). To understand the importance of this issue, it must be discriminated between the two weeding zones of the interrow and intrarow area. Several authors have noted that interrow weeding is not considered an issue but the mechanical removal of weeds from the intrarow area can still be a challenging task (Cloutier *et al.*, 2007; Tillett *et al.*, 2008; van der Linden *et al.*, 2008; van Der Weide *et al.*, 2008). Melander *et al.* (2006) described this problem with apt words: "*Interrow weeds can be removed by ordinary interrow cultivation relatively easily although intrarow weeds constitute a major challenge*". Even though the intrarow area is only a small band, comprised of the crop row width, it is especially important for weed control. Here, weeds are growing close to crop plants and the struggle for resources is higher than with weeds which grow further away in the interrow area (Zimdahl, 2004, p. 7). Therefore, targeting the intrarow weeds with suitable tools is a critical step towards effective mechanical weed control.

Another issue adding to the limited time frame to perform mechanical weeding is the high dependence on suitable soil and weather conditions. Furthermore, there is a great variety of implements for inter- and intrarow weeding. Mechanical weeding with a harrow or a hoe is preferably performed during a dry weather period and when the soil is friable. Some implements like horizontal brush hoes can also cope with lightly wet soils (Merfield, 2019). Nevertheless, wet soil conditions are not ideal for mechanical weeding because uprooted weed plants could easily regain vigor and regrow (Vasileiadis *et al.*, 2016). Other drawbacks of mechanical weeding can include the unintended propagation of perennial weeds and causing dormant seeds to be brought up to the soil surface where they germinate (Cioni and Maines, 2010). Additionally, mechanical treatments can lead to erosion due to damaging the soil structure and reduced frost tolerance of young crop plants (Kruidhof, Bastiaans and Kropff, 2008; Smith *et al.*, 2011). Especially maize is prone to frost damage if post-emergence harrowing is performed in early growth stages.

These implications highlight the complex process of mechanical weed control. Some parameters cannot be changed by the farmer. So far, this includes adjusting the treatment intensity to the crop growth stage or avoiding wet soil and unfavorable weather conditions. However, there are several options to facilitate the work and to improve the weed control result. For example, by increasing the selectivity of the mechanical weeding operation. Rasmussen *et al.* (2008) defined selectivity as the relationship between weed control and crop damage. Thus, a successful weed control treatment shows a high weed control efficacy without any crop damage. However, there is a direct possibility of damaging the crop by letting the mechanical tools injure the crop plants or indirectly by covering crop plants with high amounts of soil. By increasing the precision of the tool guidance, at least the possibility of directly injuring crop plants could be avoided. The issue of excessive soil throw can be regulated separately through adjustments of the tractor driving speed or by choosing suitable cultivation sweeps (e.g. during interrow hoeing). The increased selectivity ultimately leads to a higher weed control efficacy because the safety distance towards the crop plants can be decreased and the treated field area is increased. The uncultivated field area should be as small as possible because the benefit of mechanical weeding increases with every additional centimeter of cultivated intrarow space Cloutier *et al.* (2007). Especially in organic farming, where herbicide use is prohibited an improved weed control efficacy would be beneficial. Some organic crops like onions and other vegetables still require hand-weeding (McErlich and Boydston, 2014; Johnson *et al.*, 2017). However, hand-weeding is time-consuming and cost-intensive (Bakker *et al.*, 2010). If hand-weeding could be replaced by a precise and efficient mechanical weed control measure it would reduce the financial burden of the farmer (Melander, Rasmussen and Bàrberi, 2006).

Since the guidance precision of mechanical weed control implements like a hoe is always limited to the skills of the driver, such systems profit from sophisticated machine guidance and weed detection technology through sensors (Bond and Grundy, 2001). A wide range of sensors are suitable for the agricultural sector. However, the main applications can be found in camera guidance, laser and ultrasonic sensors as well as the global positioning system (GPS). Modern implements for mechanical weeding can be equipped with these sensors in order to perform better compared to manual steering. Based on the information gathered by the sensor(s), the implement can be steered to either follow crop rows or even recognize single crop plants (Tillett *et al.*, 2008). The sensor information is then used to actuate a steering mechanism on the implement which is usually some form of a hydraulic cylinder. However, there are also systems like the Steketee IC (Lemken GmbH & Co KG, Germany) or the Robovator (F. Poulsen Engineering ApS, Denmark) which use a pneumatic system to control the mechanical tools for hoeing.

The review study included in this thesis showed that camera-guidance systems for hoeing have proven to be effective in many arable row crops (Wiltshire, Tillett and Hague, 2003; Kunz, Weber and Gerhards, 2015; Kunz *et al.*, 2016, 2018). Their potential has been explored since the 1980s (Guyer *et al.*, 1986; Reid and Searcy, 1988) and has continued from thereon. The camera facilitates the work by increasing the working precision of the hoe. This means that the safety distance of the implements used towards the crop plants can be decreased. This, in turn, increases the weeded field area. Simultaneously, higher driving speeds can be realized with a camera-guided hoeing system without a loss of precision which leads to a higher area coverage compared to manual steering of the implement. However, the driving speed cannot be increased endlessly because at a certain speed, and depending on the used implement, the selectivity will decrease due to excessive soil movement. Typical driving speeds for interrow hoeing in wide row crops such as sugar beet, maize or soybean are in the range of 4 to 8 km h^{-1}, depending on the crop growth stage. Due to the great success of camera-guidance systems in combination with hoes, they are readily available by many manufacturers. Further advantages compared to manual steering of a hoe include less driver fatigue and that only one vehicle operator is required. The detection of single crop plants has also been realized to perform mechanical intrarow weed control (Tillett *et al.*, 2008). The tools used in such a scenario are usually two knife-like blades which open at the position of a crop plant and close after the crop plant has been passed. However, this means that adequate space between individual crop plants is necessary to perform this kind of weed control. Otherwise the tools could not work safely because of damaging neighboring or the following crop plants. Gerrish *et al.* (1997) point out that camera-guidance systems need a robust algorithm for crop row recognition. This helps the implement to stay aligned with the crop rows even if part of a crop row is missing. In order to improve the reliability of such systems further, the camera should be able to monitor more than just one crop row. In addition to this, implementing a total of two cameras, one on each side of the implement, could help to maintain a high level of row recognition in case some row parts are missing or if high weed densities make it hard for one of the cameras to recognize the crop rows (Slaughter *et al.*, 1999).

Compared to the number of studies about camera-guided systems, only a few studies dealt with laser or ultrasonic sensors for the detection of weeds or as a crop row guidance system in combination with mechanical weeding. Van Zuydam *et al.* (1995) performed hoeing in sugar beet with an implement that was guided by a laser. Another laser-guided implement was devised by Cordill and Grift (2011) for guidance of an implement along maize stalks. However, the authors stated that 23.7% of crop plants were damaged during the experiment. This system is, therefore, seen as not very efficient and reliable. Rueda-Ayala *et al.* (2015) made use of an ultrasonic sensor to estimate the weed density in maize. Depending on the measured density, the harrowing intensity was adjusted by changing the angle of the harrow tines. This kind of selective harrowing can help to minimize crop damage because regions with a low weed density are not treated as intense as regions with a high weed density.

Laser and ultrasonic sensors are well suited for equipping a vehicle with an efficient and cheap guidance system. However, camera-guidance is seen as the superior method for fast and reliable crop row recognition. The appropriate work environment for laser and ultrasonic sensors is the area of agricultural robots (autonomous systems). Since a robot is supposed to move on its own across a field it requires certain obstacle detection and avoidance systems which are usually comprised of laser or ultrasonic sensors. Lasers can help a robot to navigate across a field by letting the robot find the crop rows, but this method requires the crop to have a certain height (Reiser *et al.*, 2016). Apart from laser and ultrasonic sensors, GPS is also used mainly as a navigation method for tractors and robots. In combination with a local real-time-kinematic (RTK) reference station on the farm, sub-centimeter-level accuracy can be achieved (Hiremath *et al.*, 2013). But in general, using GPS-guidance for mechanical weed control is too inaccurate to safely perform mechanical weeding (Rasmussen *et al.*, 2012). Though, Slaughter *et al.* (2008) stated that it could be used for row guidance if the seeding has also been performed with RTK-GPS. For example, a tractor could use RTK-GPS for seeding and record the seed lines on which it traveled. These seed lines can later be used as a priori knowledge for guidance of the hoe or the tractor itself. However, from the experience of the field trials conducted for this dissertation, it was observed that the position of the seeder is not always the same. Sometimes crop rows might deviate from the ideal seeding line because of sloping terrain or because the seeder did not work accurately. In this case having a camera guidance is advantageous because it presents an online decision system tracking the crop rows no matter how they were seeded. Obviously, this only applies to row-dependent treatments like interrow hoeing. Harrowing would not require exact guidance behind the tractor because the entire field area is treated. It was found that an ideal combination for mechanical weed control is composed of RTK-GPS-steering of the tractor and a camera-guidance system on the hoe. With the RTK-GPS-steering, the tractor can drive autonomously, and the driver can cross-check the performance of the hoe behind the tractor without causing steering mistakes. The advantages of such a set-up have also been addressed by Khadraoui *et al.* (1998), stating that automatic control of speed and steering allows the operator to optimize use of the machine during the current agricultural task.

Apart from half-autonomous guidance systems, the application of agricultural robots is continually developing. These autonomous systems are supposed to perform plant protection measures without human intervention. They could help to replace repetitive and labor-intensive tasks such as hand-weeding, decrease herbicide input by substituting chemical with mechanical treatments and, due to their smaller size, reduce soil compaction (Grift *et al.*, 2008; Bawden *et al.*, 2017; Röös *et al.*, 2018). Different approaches are being tested and the employed weed control methods include chemical, mechanical or thermal weeding. For example, the Blue River Company (California, USA) developed a robot which applies tiny amounts of herbicides to individual weed plants. A similar concept for autonomous chemical weed control has been developed by ecorobotix (Switzerland) with their solar-powered robot. For mechanical weed control, Bosch Deepfield robotics uses a robot equipped with a stamping tool which pushes down on a weed, pressing it into the ground (Langsenkamp *et al.*,

2014). The robot developed by FarmWise (California, USA) also uses machine vision to distinguish weeds from crop plants but weeds are removed mechanically without using herbicides. Mechanical weed control is also the concept of the robots produced by Naïo Technologies (France) for field, greenhouse or vineyard applications. Another approach to robotic weeding is the use of a fleet of robots (multi-robot coordination) to perform persistent autonomous weed control tasks (McAllister *et al.*, 2019).

The biggest issue that robots have to overcome in an agricultural environment are the diverse field conditions: "*Complexity increases when dealing with natural objects due to the high variability of many of the parameters that affect robot behavior, many of which cannot be determined a priori*" (Bechar and Vigneault, 2017). Mostly machine vision is employed for crop row, crop plant and weed plant recognition. Hereby, weed detection presents two scenarios: the detection and removal of weeds where only soil and weed plants are present in the images of the vision system. This is commonly the case in the interrow area between two crop rows. The other scenario is to distinguish between crop and weed plants that are growing in a mixture inside the crop row (Piron, van der Heijden and Destain, 2011). This requires a robust machine vision algorithm that can cope with the highly variable field conditions due to changes in abiotic factors like ambient light or different cropping systems with varying crop plant species and appearances (Nof, 2009; Hiremath *et al.*, 2013; Bechar and Vigneault, 2017). In the future, robots will become especially important for specialty crops like flowers, herbs, and vegetables where either no or only a limited number of herbicides are permitted (Fennimore *et al.*, 2016). Such crops require intensive mechanical weed control techniques or hand-weeding. Substituting hand-weeding with an autonomous weed control system could dramatically reduce the labor-costs. Certainly, robots are quite expensive, but their advantage is that they can work continuously day and night. Also, with an increase of robots sold and used their prices are going to decrease at some point.

The second paper "**Adjustment of weed hoeing to narrowly spaced cereals**" deals with the employment of hoeing in narrow cereal row widths. Initially, the focus of sensor-guided mechanical weed control rested largely on wide row crops like sugar beet, maize and soybean where crop rows can be tracked easily. The row spacings in these crops are usually between 45 cm and 75 cm. Thus, the space to employ the cultivation sweeps is fairly large. Even though the camera steering technology has been improved during the past years, interrow hoeing still requires that the crop is seeded with a minimum row spacing of 25 cm (Hofstee and Nieuwenhuizen, 2014). However, the Austrian company Einböck claims that their hoe (model "Chopstar") in combination with the Claas camera (model "CULTI CAM", Class E-Systems GmbH, Germany) can track crop rows with a row spacing of 20 to 30 cm. Still, row spaces below 20 cm usually require the use of a harrow (Mertens and Jansen, 2002). Therefore, in German organic cereal production systems the row spacing usually ranges between 20 and 40 cm to allow interrow hoeing. However, studies have shown that increasing the row spacing can reduce grain yields by up to 16% (Boström, Anderson and Wallenhammar, 2012). Based on their findings, Fahad *et al.* (2015) also stated that narrow row distances of 11 and 15 cm resulted in higher wheat grain yields compared to a wider row spacing of 23 cm. Kirkland (1993) also recorded lower yield results for wide row spacings compared to narrow row spaces in spring barley. Due to narrow row spacings, less field area is left uncovered by the crop plants. This leads to a better weed suppression by the crop (Mohler, 2001; Kolb, Gallandt and Mallory, 2012; Gallandt *et al.*, 2015). Similar observations have been made by Teasdale and Frank (1983) in beans (*Phaseolus vulgaris* L.). It was proposed that the higher weed suppression of *P. vulgaris* sown in narrow row spacings was due to a higher rate of bean canopy closure. Gunsolus (1990) agrees with these findings and states that narrow rows allow earlier crop canopy closure, thus limiting weed emergence. Hoeing is also more aggressive than harrowing and is able to cut or uproot even large weed plants (Merfield, 2019). Additionally, hoeing could be implemented as part of an IWM strategy to further reduce the input of herbicides in conventional farms where cereal row spacings of less than 20 cm are typical. This would have the advantage of not having to change the seeding technique. Hoeing can also present an opportunity for conventional farms to regain control over an existing issue with herbicide-resistant weeds. Therefore, it would be beneficial if hoeing, especially camera-guided hoeing, could be performed in narrow row spaces. Manually guiding a hoe within 12.5 cm or 15 cm rows is tiresome for the driver and could only be performed at a slow driving speed, which usually leads to an inadequate weed control result. Therefore, an option to allow interrow hoeing in narrowly spaced cereals would be the implementation of a camera-guidance system. However, prior to developing and adapting such a guidance system to narrow row widths, it was necessary to test the effects of hoeing in narrow row spaces on the weed control efficacy, crop and weed biomass as well as the crop yield.

Thus, three different cultivation sweeps were tested in spring cereals with 12.5 and 15 cm row spacings in this thesis. At the time of the experiments, only a limited amount of information existed about the mechanical weed control in narrow row spaces. Studies were mainly concerned with harrowing instead of interrow hoeing. One of the studies concerned with hoeing was performed by Lötjönen and Mikkola (2000). The authors tested harrowing, interrow hoeing and rotary hoeing in cereals. Interrow hoeing with ducksfoot (goosefoot) sweeps was successfully performed in 18 and 25 cm spaced cereal rows. The authors stated that a slightly lower barley yield was recorded for the 25 cm space compared to the 18 cm spacing. However, the average barley yields of the interrow hoeing treatments did not differ significantly from the conventional herbicide application. Rasmussen *et al.* (2008) assessed the effect of post-emergence harrowing on the weeding selectivity in narrowly spaced cereals with row spaces of 5.3, 12 and 24 cm. Kolb *et al.* (2012) harrowed in wheat sown in three row spacings of 11, 18 and 23 cm. Since harrowing is treating the inter- and intrarow area of a crop, row guidance of the harrow did not play a role. Therefore, one objective of the experiments conducted for this thesis was to determine a suitable cultivation sweep for interrow hoeing in narrow cereal row spacings. The second objective was to analyze the effect of interrow hoeing on the crop yield and the weed control efficacy.

The weed control efficacies ranged from 34% (Kleinhohenheim) to 89% (Ihinger Hof). In 2018, Melander *et al.* (2018) also performed experiments with interrow hoeing in different row spaces. The authors found interrow hoeing to reduce the weed density by 80% to 90% in barley and by 63 to 80% in wheat. The conducted field trials for this dissertation revealed that the no-till sweeps performed well at both trial sites in terms of weed control efficacy and their performance inside each plot. An observed disadvantage of the goosefoot sweeps was the high amount of soil that was moved into the crop row during the treatments. Pullen and Cowell (1997) also stated that goosefoot (or ducksfoot) sweeps move high amounts of soil which can be a disadvantage in narrow rows because of intense crop soil burial. This can be an issue if driving speeds are increased even further. The down-cut side knives also moved high amounts of soil due to their shape. Sometimes clogging occurred, especially in the 12.5 cm row spacing. This and the fact that two down-cut side knives per interrow space had to be used makes them an unattractive choice. The results indicated slight yield differences between the different sweep types. However, none of the sweep types affected grain yields adversely and the yields were within the range of German conventional farming systems with herbicides. With the experiments it was shown that hoeing in narrow cereal row widths is not an issue from an agronomic perspective. The practical use of this interrow hoeing technique now depends on the development of a reliable machine vision algorithm to allow camera control of the hoeing implement. An implication is presented by the time frame for the application because the canopy of narrow cereal rows closes earlier than in wide cereal rows. Thus, the time to recognize crop row features is quite small and the application of interrow hoeing is going to concentrate within a short time frame.

Even though interrow hoeing can present an additional measure for mechanical weed control in narrowly spaced cereals instead of just harrowing, the importance of general farm management techniques to maintain a low weed pressure must be stressed. This includes preventive and cultural measures. Cultural measures can include a crop variety which quickly establishes a dense canopy and produces shading to suppress weed growth (Drews, Neuhoff and Köpke, 2009). The time of sowing can also have an influence on weed emergence. For example, blackgrass (*Alopecurus myosuroides* Huds.) or loose silky-bent (*Apera spica-venti* L.) germination is favored if cereals are sown in early autumn (Moss and Clarke, 1994). Other preventive practices to impede the built-up of a high weed population include crop rotation, false seedbed preparation (Dierauer and Stöppler-Zimmer, 1994; Rasmussen, 2004) and cover cropping (Teasdale, 1996). Furthermore, it is important to maintain a high level of hygiene to prevent the spread of weed seeds with machines during harvest, during soil tillage operations and to avoid the contamination of harvested crop seeds with weed seeds (Wiles and Brodahl, 2004; Benvenuti, 2007; Heijting, Van Der Werf and Kropff, 2009).

The third scientific article "**Sensor-based intrarow mechanical weed control in sugar beets with motorized finger weeders**" describes the development of a new weeding tool for mechanical intrarow weed control in sugar beet (*Beta vulgaris* subsp. *vulgaris,* Altissima Group). Alternative weed control measures are becoming increasingly important in the European sugar beet production where up to three post-emergent herbicide applications are common (Cooke and Scott, 1993, p. 507). Sugar beets are highly sensitive to resource competition through weeds (Dawson, 1965). Therefore, early weed control starting from sowing until row closure is important. Sometimes hoeing can be used to replace a late herbicide application if weed densities are low or if herbicides have been sprayed in bands over the row (Cioni and Maines, 2010). However, there is generally a high reliance on herbicides in conventional sugar beet production. Due to the abolition of the European quota system for sugar in 2017, sugar beet production increased further across the European Union (Burrell *et al.*, 2014; Götze *et al.*, 2017). However, concerns over the environmental impact of herbicides are growing because of residues that have been found in ground and surface water (Chikowo *et al.*, 2009). The development of herbicide resistance in weeds is another major factor that is redirecting weed control towards alternative weed control solutions (Buhler, 1996; Richter, Zwerger and Böttcher, 2002). This stresses the importance of improving or finding new alternative weed control techniques in sugar beets and other arable crops.

Sugar beets are cultivated in wide row spacings of 0.45 to .50 m. Thus, mechanical weed control must address the intra- and the interrow area. As with other arable crops, the removal of intrarow weeds is of great importance due to the direct resource competition (Petersen, 2004; Cloutier *et al.*, 2007; Tillett *et al.*, 2008). Already in the early 2000s, Nørremark and Griepentrog (2004) stated that there is a need to develop a novel physical selective weed control method for wide row crops like sugar beet, maize and vegetables if herbicide usage should be replaced or reduced. Therefore, the objectives of the trials conducted for this dissertation were to develop and test a new sensor-based tool to improve the mechanical intrarow weed control efficacy in wide row crops such as sugar beets. The new weeding tool was tested in a two-year sugar beet field trial. It was compared to conventional mechanical and chemical weed control measures in terms of weed control efficacy and sugar beet yield. The concept of the new tool was based on a conventional finger weeder. However, the conventional finger weeder was modified and equipped with an electric motor. This enabled the rotational movement of the finger weeder independently of the tractor driving speed. The newly developed motorized finger weeders are supposed to be used in combination with a robot to improve future mechanical weeding operations with autonomous systems.

Robots are moving quite slowly compared to a tractor. This is due to the requirement that robotic weeding needs to obtain a high weeding accuracy especially in the close-to-crop area which can momentarily only be achieved with low operational speeds of around 1 m s^{-1} or even a stop-and-go mode (Griepentrog *et al.*, 2004). Early studies using machine vision with a camera (Lee, Slaughter and Giles, 1999) still complained about slow image processing times as a limiting factor. However, today this should not be an issue anymore. But if a robot is supposed to perform mechanical weed control around single crop plants, the time for the robot to carefully employ its mechanical tools must be considered. Apart from that, the agricultural terrain itself is another issue. Khadraoui *et al.* (1998) compared the guidance of an agricultural machine to driving an off-road vehicle rather than steering a high-speed vehicle along a motorway. Examples for completely autonomous robots (not half-autonomous, tractor-pulled implements) and their corresponding speed include: the work by McAllister *et al.* (2019) who report their Agbots were moving at a speed of 3.6 km h^{-1},

Grimstad and From (2017) state their Thorvald II robot moved at 5.4 km h^{-1}; the precision spray robot by Lee *et al.* (1999) moved with 0.8 km h^{-1}; the robot by Michaels *et al.* (2015) was equipped with a stamp tool for mechanical weeding and worked at a speed of 0.13 km h^{-1}; Takai *et al.* (2014) report the use of a robot tractor moving with 2.2 km h^{-1}; the Naïo Technologies robot model "Oz" travels with a speed of 1.3 km h^{-1} and the model "Dino" reaches a maximum speed of 4 km h^{-1} (*Autonomous weeding; agricultural robots - Naïo Technologies*). However, conventional mechanical tools require a certain amount of forward travel speed to effectively cut, uproot or bury weed plants. For example, a rolling cultivator performs best between 8 to 16 km h^{-1}, rotary hoeing requires 8 to 24 km h^{-1}, harrowing is usually performed at 6 to 12 km h^{-1} and interrow hoeing requires speeds of at least 4 to 12 km h^{-1} (Bowman and Outreach, 2002). This means conventional mechanical weed control tools cannot simply be mounted onto a robot to perform mechanical weeding, especially if the beneficial effect of soil throw into the crop row wants to be preserved. From the examples of robotic working speeds, even the "faster moving" robots like the Thorvald II and Naïo's Dino are working at the lower end of the speed spectrum required for mechanical weeding. Therefore, it was proposed that only an independently rotating tool can substitute for the slower working speed of a robot as it was also used in other works including McAllister *et al.* (2019) and Åstrand and Baerveldt (2002).

Controlling interrow weeds only requires the recognition of a row structure whereas intrarow weeding requires the challenging task of recognizing individual sugar beet plants among weeds (Åstrand and Baerveldt, 2002). Therefore, a sensor system consisting of a bi-spectral camera was used for the sugar beet field trials of this dissertation. The bi-spectral camera acquired images in the red-infrared spectrum and was used in combination with an autonomously guided hoe for interrow weeding (Garford Farm Machinery Ltd., Peterborough, England). Based on the information provided by the bi-spectral camera and the software IMPASS, the objective was to apply two different rotational speeds of the motorized finger weeders. With the software IMPASS, a laptop computer and a Raspberry Pi computer thresholding was performed to produce differential images where only plant material was displayed as white and non-plant material (background, soil, stones) was depicted as black. The classification of the identified plant into the classes "weed" or "sugar beet" was then performed with the IMPASS software algorithm. The algorithm was trained with images of sugar beet and weed plants at typical development stages when mechanical weed control is performed. The IMPASS software algorithm worked reliably, however, the tractor driving speed of the operation was only 1 km h^{-1}. Thus, the speed was similar to other robotic applications of other studies performing mechanical weeding with a robot. Another issue that might arise is that the algorithm could have an issue if several plant species have to be distinguished as opposed to just discriminating between sugar beet and no sugar beet (weeds). More robust, reliable and even faster methods would include the use of convolutional neural networks as used by Dyrmann, Karstoft and Midtiby (2016). Convolutional neural networks are trained with a vast number of images and can even be used to detect weeds which are partly occluded by crop plant leaves (Dyrmann, Jørgensen and Midtiby, 2017). However, this method is more applicable for chemical weed control where a weed-species-specific herbicide is applied to individual weed plants. For mechanical weed control, it usually suffices to only detect the crop plants because the area surrounding each crop plant should be treated with the mechanical tools anyways. Soil movement induces weed germination. However, since mechanical weeding is performed at least two to three times, newly emerging weeds should not be an issue. It has been shown that even sugar beet plants can suppress weed growth past a certain developmental stage (Dawson, 1965; Dawson, 1977; Scott, Wilcockson and Moisey, 1979). Therefore, if possible, no soil region should be left untreated due to the benefits that come with every additional centimeter of cultivated space in mechanical weed control operations (Cloutier *et al.*, 2007).

Cultivating crops, especially sensitive crops like sugar beets, is a challenging task if it is performed without the use of herbicides. Therefore, achieving economically acceptable weed control and crop yields in organic cropping systems is far more challenging without herbicides (Datta & Knezevic, 2013). This can be confirmed by the results of this study. In 2017, the weed pressure was much higher compared to 2018. The mechanical treatments of the motorized finger weeders and the conventional finger weeders without an initial herbicide application performed poorly. There was still a high weed control efficacy of the mechanical treatments, ranging from 77% (conventional finger weeder) to 91% (motorized finger weeder). However, the sugar beet yields for those treatments were rather low because not all weed plants were removed from the intrarow area. Thus, these trials also demonstrate the importance of preventive measures to impede the built-up of a high weed density prior to seeding. Still, other studies reported that mechanical intrarow treatments can be as effective as the conventional herbicide application (Wiltshire, Tillett and Hague, 2003; Kunz, Weber and Gerhards, 2015). Riemens *et al.* (2007) reported weed control efficacies of up to 99% by using finger or torsion weeders. This can also be confirmed with the intrarow weed control efficacy results for the mechanical treatments obtained in 2018. However, weed densities must be at a manageable level. If weed densities appear to be too high, an initial herbicide application could be beneficial (Mulder and Doll, 1993). Replacing one out of three post-emergence herbicide applications with mechanical treatments would at least decrease herbicide usage by two thirds. Spraying herbicides only on the crop rows would reduce the herbicide input even further. Overall, it can be concluded that the motorized finger weeders in combination with the bi-spectral camera and the autonomous row guidance for interrow hoeing performed well. Motorized finger weeders are seen as a promising solution to perform mechanical intrarow weed control in future weeding operations. They are the ideal tool for an autonomous system (robot) and could help to further reduce herbicide inputs in arable crops.

Modern IWM approaches have revived mechanical weed control techniques and moved them from a purely organic perspective to conventional cropping systems. This work analyzed the developments in modern sensor-based mechanical weed control and assessed how they could be expanded further. However, mechanical weed control requires broad agricultural knowledge and the complexity of the topic might deter some farmers (Chikowo, 2009; Merfield, 2019). Most importantly it must be realized that mechanical weed control cannot be a stand-alone solution to the weed problem. Davies (2002) and Bàrbieri (2003) describe weed control as a combination of preventive and curative weed control measures. Preventive measures can include crop rotation, false seedbed preparation, time of sowing and seed quality whereas curative (reactive) measures are some form of mechanical intervention in the growing crop like hoeing or harrowing. Since the weed problem cannot be solved by curative tactics alone, the aim should be to maintain a balanced system where the development of large weed populations is suppressed with every possible means throughout the year and should involve the whole cropping system (Liebman & Davies, 2000; Bond & Grundy, 2001; Kruidhof *et al.*, 2008). Mechanical weed control will always be the tip of the iceberg in a long chain of preventive and reactive measurements to keep the overall weed density low. Therefore, believing that mechanical weed control, whether with or without sensor-assistance can turn a neglected field into a weed-free environment must be considered wishful thinking. Agriculture is not simply the improvement of one process alone. A holistic approach considering every possible aspect must be the aim. However, it can be concluded that sensor-supported machinery, especially for mechanical weeding, can help the farmer to facilitate the labor and increase the efficiency of weed control operations.

4 SUMMARY

Weeds, apart from insect pests and pathogens, are the major reason for yield losses. Weeds in arable cropping systems are most commonly managed by the application of herbicides. However, several weed species have developed herbicide resistance. Simultaneously the concern about the impact of herbicides on the environment and public health increased. Using other weed control measures like mechanical weeding can be an option to reduce the input of herbicides in agriculture. But alternative weeding methods must also be able to compete with herbicide performance. The implementation of sensor technology in mechanical weed control can help to improve the overall performance of mechanical weeding. The benefits of sensor-assisted mechanical weed control implements include an increase of weed control efficacy due to a higher precision of the tool guidance which ultimately results in an increased weeding selectivity and a higher working speed. The combination of a hoe with a camera guidance system has already been established in numerous row crops such as sugar beet, soybean and maize. The camera guidance system helps to guide the hoe along the crop rows whereby crop plant damage is minimized. However, hoeing is usually only performed in wide row spacings and it has not yet been tested in conventional narrow row spaces of less than 20 cm. Other sensor types, including laser and ultrasonic sensors, are mainly used in robotic systems that perform autonomous fieldwork and require several obstacle avoidance systems. Using robotic systems for weeding is challenging compared to standardized industrial applications because a field is a diverse environment, and this complicates the autonomous weed control tasks. Mechanical weeding is a multifaceted area of agriculture and requires high expertise, especially in combination with state-of-the-art technology.

Therefore, this dissertation assessed several aspects of modern mechanical weed control including i) a review of the latest developments in sensor-based mechanical weed control, ii) the possibility of hoeing in conventional narrow row spaces and iii) the development and testing of a new implement for mechanical intrarow weed control with an autonomous system. The following objectives were addressed within this thesis:

Review of current developments in sensor technology for mechanical weed control and estimating future developments;

Evaluation of hoeing in conventional narrow cereal row spaces and the effect on weed control efficacy and yield in comparison to conventional herbicide applications;

Identifying a suitable tool for mechanical weed control in conventional narrow cereal row spaces at various driving speeds;

Developing and testing a new implement intended for sensor-based intrarow mechanical weed control with autonomous systems in sugar beet.

The research objectives were addressed in three scientific publications. The **first publication** reviews the developments of sensor-guided mechanical weed control. Autonomous systems (robots) have only recently been introduced to the agricultural sector. However, camera-guided implements for mechanical weed control in row crops have been available on the markets for several years. Camera guidance systems are currently the most advanced sensor system and further development to increase their robustness will help to cope with changing environmental conditions in the fields. Laser and ultrasonic sensors and GPS-guidance via satellite are mainly used to assist with the guidance of a tractor or a robot inside the field. They are rarely used for direct interactions between a weeding tool and a weed plant. Robotic field applications are expected to increase in areas with exhausting and time-consuming tasks such as mechanical weeding.

The **second paper** discusses the feasibility of hoeing in conventional narrow cereal row spaces of 12.5 and 15 cm. Two field experiments were conducted in southwest Germany to assess the effects of hoeing with goosefoot sweeps, no-till sweeps and down-cut side knives in spring barley (*Hordeum vulgare* L.) and spring oats (*Avena sativa* L.). Interrow hoeing was successful with all three cultivation sweeps and no significant crop yield losses occurred compared to treatments with a conventional herbicide application or manual weeding. The weed control efficacies ranged between 51 and 89%. No interaction was found between the tractor driving speeds of 4 and 6 km h^{-1}, the sweep type and their effect on the weed control efficacy. The no-till sweeps appeared to be the most promising tool design for hoeing in narrowly spaced cereals out of the evaluated sweeps. They cut through the soil easily without excessive soil movement. This is an important feature allowing the employment of no-till sweeps at driving speeds above 6 km h^{-1} without crop damage due to increased soil throw. Interrow hoeing in narrow row spaces proved to be feasible and combining this method with an automatic guidance system can become a considerable option to reduce the herbicide input.

The third **scientific article** is concerned with sensor-guided intra- and interrow mechanical weed control in sugar beet. A new intrarow weeding tool was developed and tested in combination with an autonomous crop row guidance system. The new weeding tool was based on a conventional finger weeder that was equipped with an electric motor. This enabled the rotation of the finger weeders independent of the forward travel speed of the tractor. Two different rotational speeds were tested and compared to the weed control efficacy of conventional finger weeders and conventional herbicide applications. The motorized finger weeders reached intrarow weed control efficacies between 87 and 94%. Therefore, the motorized finger weeders are a promising tool design for intrarow mechanical weed control for autonomous weeding applications.

This dissertation assessed the importance and possibilities of sensor-based mechanical weed control as an alternative weeding method to conventional herbicide application in arable crops. Modern mechanical weed control must keep pace with technological advancements. This demands the continuous refinement of existing machinery and the development of new weeding technologies and tools to constantly improve the weed control performance of mechanical weeding operations. It was shown that hoeing is a suitable option for mechanical weed control even in narrowly spaced crops. Furthermore, a new concept for sensor-based intrarow mechanical weed control in sugar beets was successfully developed. These findings contribute to modern plant protection strategies by expanding the possibilities of mechanical weed control to further reduce the reliance on herbicides in arable crops.

5 ZUSAMMENFASSUNG

Unkräuter sind neben Insektenschädlingen und Krankheitserregern die Hauptursache für Ertragseinbußen. In Ackerbausystemen werden Unkräuter größtenteils durch den Einsatz von Herbiziden bekämpft. Allerdings haben mehrere Unkrautarten eine Herbizidresistenz entwickelt. Gleichzeitig nahm die Sorge um die Auswirkungen von Herbiziden auf die Umwelt und die öffentliche Gesundheit zu. Die Verwendung anderer Unkrautbekämpfungsmaßnahmen wie der mechanischen Unkrautbekämpfung kann eine Option sein, um den Herbizideinsatz in der Landwirtschaft zu reduzieren. Aber auch alternative Unkrautbekämpfungsmethoden müssen mit Herbiziden auf einer wirtschaftlichen Ebene konkurrieren können. Der Einsatz von Sensortechnik in der mechanischen Unkrautbekämpfung kann dazu beitragen die Gesamtleistung der mechanischen Unkrautbekämpfung zu verbessern. Die Vorteile der sensorgestützten mechanischen Unkrautbekämpfung liegen in der Steigerung der Unkrautbekämpfung durch eine höhere Präzision der Werkzeugführung, was letztendlich zu einer erhöhten Unkrautselektivität und einer höheren Arbeitsgeschwindigkeit führt. Die Kombination einer Hacke mit einem Kameralenksystem hat sich bereits in zahlreichen Reihenkulturen wie Zuckerrüben, Sojabohnen und Mais etabliert. Das Kameralenksystem hilft, die Hacke entlang der Pflanzenreihen zu führen, wodurch die Schäden an der Kulturpflanze minimiert werden. Das Hacken wird jedoch in der Regel nur in weiten Reihenabständen durchgeführt und ist noch nicht in konventionellen engen Reihenabständen von weniger als 20 cm getestet worden. Andere Sensortypen, einschließlich Laser- und Ultraschallsensoren, werden hauptsächlich in Robotersystemen eingesetzt, die autonome Feldarbeit leisten und mehrere Hinderniserkennungssysteme erfordern. Der Einsatz von Robotersystemen zur Unkrautbekämpfung ist im Vergleich zu standardisierten industriellen Anwendungen eine Herausforderung, da ein Feld eine abwechslungsreiche Struktur besitzt und autonome Unkrautbekämpfungsaufgaben deutlich erschwert. Die mechanische Unkrautbekämpfung ist ein vielschichtiger Bereich der Landwirtschaft und erfordert eine hohe Kompetenz, insbesondere in Kombination mit modernster Technik.

Daher wurden in dieser Arbeit mehrere Aspekte der modernen mechanischen Unkrautbekämpfung untersucht, darunter i) ein Überblick der Entwicklungen in der sensorgestützten mechanischen Unkrautbekämpfung, ii) die Möglichkeit des Hackens in konventionellen engen Reihenabständen und iii) die Entwicklung und Erprobung eines neuen Gerätes zur mechanischen Unkrautkontrolle in Zuckerrübenreihen mit einem autonomen System. Die folgenden Ziele wurden im Rahmen dieser Arbeit behandelt:

Untersuchung aktueller Entwicklungen in der Sensortechnik zur mechanischen Unkrautbekämpfung und Abschätzung zukünftiger Entwicklungen;

Bewertung des Hackens in konventionellen engen Getreidereihenabständen und dessen Auswirkungen auf den Unkrautbekämpfungserfolg und den Ertrag im Vergleich zur konventionellen Herbizidanwendung;

Identifizierung eines geeigneten Werkzeugs zur mechanischen Unkrautbekämpfung in konventionellen engen Getreidereihenabständen bei verschiedenen Fahrgeschwindigkeiten;

Entwicklung und Erprobung eines neuen Gerätes zur sensorgestützten mechanischen Unkrautbekämpfung in Zuckerrübenreihen

Die Forschungsziele wurden in drei wissenschaftlichen Publikationen behandelt. Die **erste Veröffentlichung** befasst sich mit einem Überblick bezüglich der Entwicklungen sensorgesteuerter mechanischer Unkrautbekämpfung. Autonome Systeme (Roboter) wurden erst kürzlich in der Landwirtschaft eingeführt. Kamerageführte Geräte zur mechanischen Unkrautbekämpfung in Reihenkulturen sind jedoch seit einigen Jahren auf den Märkten erhältlich. Kameragesteuerte Lenksysteme sind derzeit das fortschrittlichste Sensorsystem. Die Weiterentwicklung der Kameratechnik wird weiter zur Erhöhung ihrer Zuverlässigkeit beitragen, um sich noch besser an die unterschiedlichen Umweltbedingen im Feld anzupassen. Laser- und Ultraschallsensoren sowie die GPS-Führung über Satelliten werden hauptsächlich zur Unterstützung bei der Navigation eines Traktors oder Roboters auf dem Feld eingesetzt. Diese Sensoren werden selten für den direkten Einsatz eines Werkzeugs zur mechanischen Unkrautkontrolle verwendet. In Arbeitsbereichen mit anstrengenden und zeitaufwändigen Aufgaben wie der mechanischen Unkrautbekämpfung werden zunehmend robotergestützte Feldanwendungen erwartet.

Der **zweite Artikel** diskutiert die mechanische Unkrautbekämpfung durch Hacken in konventionellen engen Getreidereihenabständen von 12,5 und 15 cm. Hierzu wurden zwei Feldexperimente durchgeführt, um die Auswirkungen des Hackens mit

Gänsefußscharen, Flachhackscharen und Winkelmessern in Sommergerste (*Hordeum vulgare* L.) und Hafer (*Avena sativa* L.) zu untersuchen. Mit allen drei Scharformen konnte erfolgreich gehackt werden und es traten keine signifikanten Ertragseinbußen auf im Vergleich zu Behandlungen mit einer konventionellen Herbizidanwendung oder manueller Unkrautbekämpfung. Der Unkrautbekämpfungserfolg lag zwischen 51 und 89%. Es wurde keine Wechselwirkung zwischen der Fahrgeschwindigkeit des Traktors von 4 und 6 km h^{-1}, dem Schartyp und deren Einfluss auf die Wirksamkeit der Unkrautbekämpfung festgestellt. Die Flachhackschare erwiesen sich jedoch als das vielversprechendste Werkzeugdesign für das Hacken in engen Getreidereihen. Sie durchschnitten den Boden leichtgängig und verursachten keine übermäßig starke Bodenbewegung. Dies ermöglicht den Einsatz von Flachhackscharen auch bei Fahrgeschwindigkeiten über 6 km h^{-1} ohne Kulturpflanzenschäden durch erhöhte Bodenbewegungen zu verursachen. Das Hacken zwischen der Reihe in engen Reihenabständen hat sich als durchführbar erwiesen und eine Kombination dieser Methode mit einem automatischen Lenksystem kann zukünftig einen erheblichen Beitrag zur Reduzierung des Herbizideinsatzes in konventionellen Produktionssystemen leisten.

Der **dritte wissenschaftliche Artikel** beschäftigt sich mit der sensorgesteuerten mechanischen Unkrautbekämpfung in Zuckerrüben innerhalb und zwischen den Reihen. Ein neues Gerät zur Bekämpfung von Unkräutern in der Reihe wurde entwickelt und in Kombination mit einem autonomen Reihenerkennungssystem getestet. Das neue Werkzeug basiert auf einer herkömmlichen Fingerhacke, die mit einem Elektromotor ausgestattet wurde. Dadurch konnte sich die Fingerhacke unabhängig von der Fahrgeschwindigkeit drehen. Zwei unterschiedliche Umdrehungsgeschwindigkeiten wurden getestet und mit der Unkrautbekämpfung einer herkömmlichen Fingerhacke und einer konventionellen Herbizidapplikation verglichen. Die motorisierte Fingerhacke erreichte eine Reduktion der Unkrautdichte um 87 bis 94% innerhalb der Rübenreihen. Daher sind die motorisierten Fingerhacken ein vielversprechendes Werkzeugdesign für die autonome mechanische Unkrautkontrolle innerhalb von Kulturpflanzenreihen.

In dieser Dissertation wurden die Bedeutung und Möglichkeiten der sensorgestützten mechanischen Unkrautbekämpfung als Alternative zur konventionellen Herbizidanwendung in Ackerkulturen untersucht. Moderne mechanische Unkrautbekämpfung muss auf dem neuesten technologischen Stand sein. Dies erfordert die kontinuierliche Weiterentwicklung bestehender Maschinen und die Entwicklung neuer

Unkrautbekämpfungsmethoden, um die Effektivität der einzelnen Bekämpfungsmaßnahmen stetig zu verbessern. Es zeigte sich, dass das Hacken eine geeignete Methode für die mechanische Unkrautbekämpfung auch in konventionell gesäten engen Reihenabständen darstellt. Darüber hinaus wurde erfolgreich ein neues Konzept zur sensorgestützten Unkrautbekämpfung in Zuckerrübenreihen entwickelt. Diese Erkenntnisse tragen zu modernen Pflanzenschutzstrategien bei, indem sie die Möglichkeiten der mechanischen Unkrautbekämpfung erweitern, um den Herbizideinsatz in Ackerkulturen weiter zu reduzieren.

6 GENERAL REFERENCES

ABBAS, T., ZAHIR, Z. A., NAVEED, M. AND KREMER, R. J. (2018) 'Limitations of Existing Weed Control Practices Necessitate Development of Alternative Techniques Based on Biological Approaches', in *Advances in Agronomy*. Academic Press, pp. 239–280.

ARMENGOT, L., JOSÉ-MARÍA, L., CHAMORRO, L. AND SANS, F. X. (2013) 'Weed harrowing in organically grown cereal crops avoids yield losses without reducing weed diversity', *Agronomy for Sustainable Development*. Springer-Verlag, **33**(2), pp. 405–411.

ÅSTRAND, B. AND BAERVELDT, A.-J. (2002) 'An agricultural mobile robot with vision-based perception for mechanical weed control', **13**(1), pp. 21–35.

Autonomous weeding; agricultural robots - Naïo Technologies (2019). Available at: https://www.naio-technologies.com/en/ (Accessed: 2 August 2019).

BAKKER, T., ASSELT, K., BONTSEMA, J., MÜLLER, J. AND STRATEN, G. (2010) 'Systematic design of an autonomous platform for robotic weeding', *Journal of Terramechanics*, **47**(2), pp. 63–73.

BASTIAANS, L., PAOLINI, R. AND BAUMANN, D. T. (2008) 'Focus on ecological weed management: What is hindering adoption?', *Weed Research*, pp. 481–491.

BAWDEN, O., KULK, J., RUSSELL, R., MCCOOL, C., ENGLISH, A., DAYOUB, F., LEHNERT, C. AND PEREZ, T. (2017) 'Robot for weed species plant-specific management', *Journal of Field Robotics*, **34**(6), pp. 1179–1199.

BECHAR, A. AND VIGNEAULT, C. (2017) 'Agricultural robots for field operations. Part 2: Operations and systems', *Biosystems Engineering*, **153**, pp. 110–128.

BENVENUTI, S. (2007) 'Weed seed movement and dispersal strategies in the agricultural environment', *Weed Biology and Management*, pp. 141–157.

BOND, W. AND GRUNDY, A. C. (2001) 'Non-chemical weed management in organic farming systems', *Weed Research*, pp. 383–405.

BOSTRÖM, U., ANDERSON, L. E. AND WALLENHAMMAR, A. C. (2012) 'Seed distance in relation to row distance: Effect on grain yield and weed biomass in organically grown winter wheat, spring wheat and spring oats', *Field Crops Research*. Elsevier, **134**, pp. 144–152.

BOWMAN, G. AND OUTREACH, S. (2002) *Steel in the field: a farmer's guide to weed management tools, Sustainable Agriculture Network handbook series (USA).*

BUHLER, D. D. (1996) 'Development of alternative weed management strategies', in *Journal of Production Agriculture*. American Society of Agronomy, Crop Science Society of America, Soil Science Society of America, pp. 501–505.

BUHLER, D. D. (2002) '50th Anniversary—Invited Article: Challenges and opportunities for integrated weed management', *Weed Science*. Cambridge University Press, **50**(3), pp. 273–280.

BUHLER, D. D., LIEBMAN, M. AND OBRYCKI, J. J. (2006) 'Theoretical and practical challenges to an IPM approach to weed management', *Weed Science*, **48**(3), pp. 274–280.

BURRELL, A., HIMICS, M., VAN DOORSLAER, B., CIAIAN, P. AND SHRESTHA, S. (2014) *EU sugar policy: A sweet transition after 2015?*

CHIKOWO, R., FALOYA, V., PETIT, S. AND MUNIER-JOLAIN, N. M. (2009) 'Integrated Weed Management systems allow reduced reliance on herbicides and long-term weed control', *Agriculture, Ecosystems and Environment*, **132**(3–4), pp. 237–242.

CIONI, F. AND MAINES, G. (2010) 'Weed Control in Sugarbeet', *Sugar Tech*. Springer-Verlag, pp. 243–255.

CLOUTIER, D. C., VAN DER WEIDE, R. Y., PERUZZI, A. AND LEBLANC, M. L. (2007) '8 Mechanical Weed Management', *Non-chemical Weed management*, p. 111.

COOKE, D. AND SCOTT, J. (1993) *The Sugar Beet Crop, The Sugar Beet Crop.*

CORDILL, C. AND GRIFT, T. E. (2011) 'Design and testing of an intra-row mechanical weeding machine for corn', *Biosystems Engineering*, **110**(3), pp. 247–252.

DAWSON, J. H. (1965) 'Competition between Irrigated Sugar Beets and Annual Weeds', *Weeds*, **13**(3), p. 245.

DAWSON, J. H. (1977). Competition of late-emerging weeds with sugarbeets. *Weed Science*, **25**(2), 168-170.

DÉLYE, C., JASIENIUK, M. AND LE CORRE, V. (2013) 'Deciphering the evolution of herbicide resistance in weeds', *Trends in Genetics*, pp. 649–658.

DIERAUER, H.-U. AND STÖPPLER-ZIMMER, H. (1994) *Unkrautregulierung ohne Chemie*. Ulmer.

DREWS, S., NEUHOFF, D. AND KÖPKE, U. (2009) 'Weed suppression ability of three winter wheat varieties at different row spacing under organic farming conditions', *Weed Research*, **49**(5), pp. 526–533.

DYRMANN, M., JØRGENSEN, R. N. AND MIDTIBY, H. S. (2017) 'RoboWeedSupport - Detection of weed locations in leaf occluded cereal crops using a fully convolutional neural network', *Advances in Animal Biosciences*. Cambridge University Press, **8**(2), pp. 842–847.

DYRMANN, M., KARSTOFT, H. AND MIDTIBY, H. S. (2016) 'Plant species classification using deep convolutional neural network', *Biosystems Engineering*, **151**, pp. 72–80.

FAHAD, S., HUSSAIN, S., CHAUHAN, B. S., SAUD, S., WU, C., HASSAN, S., TANVEER, M., JAN, A. AND HUANG, J. (2015) 'Weed growth and crop yield loss in wheat as influenced by row spacing and weed emergence times', *Crop Protection*. Elsevier, **71**, pp. 101–108.

FENNIMORE, S. A., SLAUGHTER, D. C., SIEMENS, M. C., LEON, R. G. AND SABER, M. N. (2016) 'Technology for Automation of Weed Control in Specialty Crops', *Weed Technology*. Cambridge University Press, **30**(4), pp. 823–837.

GALLANDT, E. R., WEINER, J., GALLANDT, E. R. AND WEINER, J. (2015) 'Crop-Weed Competition', in *eLS*. Chichester, UK: John Wiley & Sons, Ltd, pp. 1–9.

GERRISH, J. B., FEHR, B. W., VAN EE, G. R. AND WELCH, D. P. (1997) 'Self-steering tractor guided by computer-vision', *Applied Engineering in Agriculture*, **13**(5), pp. 559–563.

GIANESSI, L. P. (2013) 'The increasing importance of herbicides in worldwide crop production', *Pest Management Science*. John Wiley & Sons, Ltd, pp. 1099–1105.

GOBOR, Z. (2013) 'Mechatronic System for Mechanical Weed Control of the Intra-row Area in Row Crops', *KI - Künstliche Intelligenz*, **27**(4), pp. 379–383.

GÖTZE, P., RÜCKNAGEL, J., WENSCH-DORENDORF, M., MÄRLÄNDER, B. AND CHRISTEN, O. (2017) 'Crop rotation effects on yield, technological quality and yield stability of sugar beet after 45 trial years', *European Journal of Agronomy*, **82**, pp. 50–59.

GRIFT, T., ZHANG, Q., KONDO, N. AND TING, K. (2008) 'A review of automation and robotics for the bio-industry', *Journal of Biomechatronics Engineering*, **1**(1), pp. 37–54.

GRIMSTAD, L. AND FROM, P. J. (2017) 'Thorvald II - a Modular and Re-configurable Agricultural Robot', *IFAC-PapersOnLine*. Elsevier, **50**(1), pp. 4588–4593.

GUNSOLUS, J. L. (1990) 'Mechanical and cultural weed control in corn and soybeans', *American Journal of Alternative Agriculture*. Cambridge University Press, **5**(3), pp. 114–119.

GUYER, D. E., MILES, G. E., SCHREIBER, M. M., MITCHELL, O. R. AND VANDERBILT, V. C. (1986) 'Machine vision and image processing for plant identification', *Transactions of the ASAE*, **29**(6), pp. 1500–1507.

HAHN, M., SCHOTTHÖFER, A., SCHMITZ, J., FRANKE, L. A. AND BRÜHL, C. A. (2015) 'The effects of agrochemicals on Lepidoptera, with a focus on moths, and their pollination service in field margin habitats', *Agriculture, Ecosystems and Environment*, **207**, pp. 153–162.

HARKER, K. N. AND O'DONOVAN, J. T. (2013) 'Recent Weed Control, Weed Management, and Integrated Weed Management', *Weed Technology*. Cambridge University Press, **27**(1), pp. 1–11.

HATCHER, P. E. AND MELANDER, B. (2003) 'Combining physical, cultural and biological methods: Prospects for integrated non-chemical weed management strategies', *Weed Research*, pp. 303–322.

HEIJTING, S., VAN DER WERF, W. AND KROPFF, M. J. (2009) 'Seed dispersal by forage harvester and rigid-tine cultivator in maize', *Weed Research*. John Wiley & Sons, Ltd (10.1111), **49**(2), pp. 153–163.

HIREMATH, S. A., STEIN, A., TER BRAAK, C. J. F., VAN DER HEIJDEN, G. W. A. M. AND VAN EVERT, F. K. (2013) 'Laser range finder model for autonomous navigation of a robot in a maize field using a particle filter', *Computers and Electronics in Agriculture*, **100**, pp. 41–50.

HOFSTEE, J. W. AND NIEUWENHUIZEN, A. T. (2014) 'Field applications of automated weed control: Northwest Europe', in *Automation: The Future of Weed Control in Cropping Systems*. Dordrecht: Springer Netherlands, pp. 171–187.

HOLZNER, W. AND NUMATA, M. (1982) *Biology and ecology of weeds*. Springer Netherlands.

HOME, M. C. W., TILLETT, N. D., HAGUE, T. AND GODWIN, R. J. (2002) 'An experimental study of lateral positional accuracy achieved during inter-row cultivation', in *Proceedings of the 5th EWRS Workshop on Physical and Cultural Weed Control. Scuola Superiore Sant'Anna di studi universitari e di perfezionamento, Pisa, Italy. 11-13 March 2002*, pp. 101–110.

International Society of Precision Agriculture (2019). Available at: https://www.ispag.org/ (Accessed: 9 September 2019).

JOHNSON, W. C., DUTTA, B., SANDERS, F. H. AND LUO, X. (2017) 'Interactions among Cultivation, Weeds, and a Biofungicide in Organic Vidalia ® Sweet Onion', *Weed Technology*. Cambridge University Press, **31**(6), pp. 890–896.

KHADRAOUI, D., DEBAIN, C., ROUVEURET, R., MARTINET, P., BONTON, P. AND GALLICE, J. (1998) 'Vision-based control in driving assistance of agricultural vehicles', *International Journal of Robotics Research*. Sage PublicationsSage CA: Thousand Oaks, CA, **17**(10), pp. 1040–1054.

KIRKLAND, K. J. (1993) 'Weed management in spring barley(hordeum vulgare)in the absence of herbicides', *Journal of Sustainable Agriculture*. Taylor & Francis Group, **3**(3–4), pp. 95–104.

KOGAN, M. (1998) 'Integrated Pest Management: Historical Perspectives and Contemporary Developments', *Annual Review of Entomology*. Annual Reviews 4139 El Camino Way, P.O. Box 10139, Palo Alto, CA 94303-0139, USA, **43**(1), pp. 243–270.

KOLB, L. N., GALLANDT, E. R. AND MALLORY, E. B. (2012) 'Impact of Spring Wheat Planting Density, Row Spacing, and Mechanical Weed Control on Yield, Grain Protein, and Economic Return in Maine', *Weed Science*. Weed Science Society of America 810 East 10th Street, Lawrence, KS 66044-8897, **60**(02), pp. 244–253.

KOUWENHOVEN, J. K. (1997) 'Intra-row mechanical weed control - Possibilities and problems', *Soil and Tillage Research*, **41**(1–2), pp. 87–104.

KOVÁCS-HOSTYÁNSZKI, A., ESPÍNDOLA, A., VANBERGEN, A. J., SETTELE, J., KREMEN, C. AND DICKS, L. V. (2017) 'Ecological intensification to mitigate impacts of conventional intensive land use on pollinators and pollination', *Ecology Letters*, pp. 673–689.

KRUIDHOF, H. M., BASTIAANS, L. AND KROPFF, M. J. (2008) 'Ecological weed management by cover cropping: Effects on weed growth in autumn and weed establishment in spring', *Weed Research*, **48**(6), pp. 492–502.

KUNZ, C., WEBER, J. F., PETEINATOS, G. G., SÖKEFELD, M. AND GERHARDS, R. (2018) 'Camera steered mechanical weed control in sugar beet, maize and soybean', *Precision Agriculture*, **19**(4), pp. 708–720.

KUNZ, C., WEBER, J. AND GERHARDS, R. (2015) 'Benefits of Precision Farming Technologies for Mechanical Weed Control in Soybean and Sugar Beet—Comparison of Precision Hoeing with Conventional Mechanical Weed Control', *Agronomy*. Multidisciplinary Digital Publishing Institute, **5**(2), pp. 130–142.

KUNZ, C., WEBER, J., JULIUS-KÜHN-ARCHIV, R. G.- AND 2016, U. (2016) 'Comparison of different mechanical weed control strategies in sugar beets', *Julius-Kühn-Archiv*, (452), p. 446.

KURSTJENS, D. A. G. AND KROPFF, M. J. (2001) 'The impact of uprooting and soil-covering on the effectiveness of weed harrowing', *Weed Research*. John Wiley & Sons, Ltd (10.1111), **41**(3), pp. 211–228.

KURSTJENS, D. A. G. AND PERDOK, U. D. (2000) 'The selective soil covering mechanism of weed harrows on sandy soil', *Soil and Tillage Research*. Elsevier, **55**(3–4), pp. 193–206.

LANGSENKAMP, F., SELLMANN, F., KOHLBRECHER, M., KIELHORN, A., STROTHMANN, W., MICHAELS, A., RUCKELSHAUSEN, A. AND TRAUTZ, D. (2014) 'Tube Stamp for mechanical intra-row individual Plant Weed Control', *Agricultural Engineering International: the CIGR Ejournal*, pp. 1–11.

LEE, W. S., SLAUGHTER, D. C. AND GILES, D. K. (1999) 'Robotic Weed Control System for Tomatoes', *Precision Agriculture*. Kluwer Academic Publishers, **1**(1), pp. 95–113.

VAN DER LINDEN, S., MOUAZEN, A. M., ANTHONIS, J., RAMON, H. AND SAEYS, W. (2008) 'Infrared laser sensor for depth measurement to improve depth control in intra-row mechanical weeding', *Biosystems Engineering*, **100**(3), pp. 309–320.

LÖTJÖNEN, T. AND MIKKOLA, H. (2000) 'Three mechanical weed control techniques in spring cereals', *Agricultural and Food Science in Finland*, **9**(4), pp. 269–278.

MCALLISTER, W., OSIPYCHEV, D., DAVIS, A. AND CHOWDHARY, G. (2019) 'Agbots: Weeding a field with a team of autonomous robots', *Computers and Electronics in Agriculture*. Elsevier, **163**, p. 104827.

MCERLICH, A. F. AND BOYDSTON, R. A. (2014) 'Current state of weed management in organic and conventional cropping systems', in *Automation: The Future of Weed Control in Cropping Systems*. Dordrecht: Springer Netherlands, pp. 11–32.

MELANDER, B., JABRAN, K., DE NOTARIS, C., ZNOVA, L., GREEN, O. AND OLESEN, J. E. (2018) 'Inter-row hoeing for weed control in organic spring cereals—Influence of inter-row spacing and nitrogen rate', *European Journal of Agronomy*. Elsevier, **101**, pp. 49–56.

MELANDER, B., RASMUSSEN, I. A. AND BÀRBERI, P. (2006) 'Integrating physical and cultural methods of weed control— examples from European research', *Weed Science*. Cambridge University Press, **53**(3), pp. 369–381.

MERFIELD, C. N. (2019) 'Integrated Weed Management in Organic Farming', in *Organic Farming*. Woodhead Publishing, pp. 117–180.

MERTENS, S. K. AND JANSEN, J. H. (2002) 'Weed seed production, crop planting pattern, and mechanical weeding in wheat', *Weed Science*. Cambridge University Press, **50**(6), pp. 748–756.

MICHAELS, A., HAUG, S. AND ALBERT, A. (2015) 'Vision-based high-speed manipulation for robotic ultra-precise weed control', in *IEEE International Conference on Intelligent Robots and Systems*. IEEE, pp. 5498–5505.

MOHLER, C. (2007) 'Mechanical Weed Control in Agriculture', in *Encyclopedia of Pest Management, Volume II*. Berlin, Heidelberg: Springer Berlin Heidelberg, pp. 338–343.

MOHLER, C. L. (2001) 'Enhancing the competitive ability of crops', in *Ecological Management of Agricultural Weeds*, pp. 269–321.

MOSS, S. R. (1993) 'Herbicide-Resistant weeds: A worldwide perspective', *The Journal of Agricultural Science*. Cambridge University Press, pp. 141–148.

MOSS, S. R. (2010) 'Integrated weed management (IWM): will it reduce herbicide use?', *Communications in agricultural and applied biological sciences*, **75**(2), pp. 9–17.

MOSS, S. R. AND CLARKE, J. H. (1994) 'Guidelines for the prevention and control of herbicide-resistant black-grass (Alopecurus myosuroides Huds.)', *Crop Protection*. Elsevier, **13**(3), pp. 230–234.

MULDER, T. A. AND DOLL, J. D. (1993) 'Integrating Reduced Herbicide Use with Mechanical Weeding in Corn (Zea mays)', *Weed Technology*. Cambridge University Press, **7**(2), pp. 382–389.

MULLA, D. J. (2013) 'Twenty five years of remote sensing in precision agriculture: Key advances and remaining knowledge gaps', *Biosystems Engineering*. Academic Press, pp. 358–371.

NOF, S. Y. (2009) *Springer handbook of automation*. Edited by S. Y. Nof. Springer Berlin Heidelberg.

NØRREMARK, M. AND GRIEPENTROG, H. W. (2004) 'Analysis and Definition of the close-to-crop Area in Relation to Robotic Weeding', *6th Workshop of EWRS Working Group Physical and Cultural Weed Control*, pp. 1–14.

OERKE, E. C. (2006) 'Crop losses to pests', *Journal of Agricultural Science*. Cambridge University Press, pp. 31–43.

ÖZKARA, A., AKYIL, D. AND KONUK, M. (2016) 'Pesticides, Environmental Pollution, and Health', in *Environmental Health Risk - Hazardous Factors to Living Species*. InTech.

PETERSEN, J. (2004) 'A Review on Weed Control in Sugarbeet', in *Weed Biology and Management*. Dordrecht: Springer Netherlands, pp. 467–483.

PIRON, A., VAN DER HEIJDEN, F. AND DESTAIN, M. F. (2011) 'Weed detection in 3D images', *Precision Agriculture*, **12**(5), pp. 607–622.

PULLEN, D. W. M. AND COWELL, P. A. (1997) 'An Evaluation of the Performance of Mechanical Weeding Mechanisms for use in High Speed Inter-Row Weeding of Arable Crops', *Journal of Agricultural Engineering Research*. Academic Press, **67**(1), pp. 27–34.

RASMUSSEN, I. A. (2004) 'The effect of sowing date, stale seedbed, row width and mechanical weed control on weeds and yields of organic winter wheat', *Weed Research*. John Wiley & Sons, Ltd (10.1111), **44**(1), pp. 12–20.

RASMUSSEN, J. (1992) 'Testing harrows for mechanical control of annual weeds in agricultural crops', *Weed Research*. Blackwell Publishing Ltd, **32**(4), pp. 267–274.

RASMUSSEN, J., BIBBY, B. M. AND SCHOU, A. P. (2008) 'Investigating the selectivity of weed harrowing with new methods', *Weed Research*. John Wiley & Sons, Ltd (10.1111), **48**(6), pp. 523–532.

RASMUSSEN, J., GRIEPENTROG, H. W., NIELSEN, J. AND HENRIKSEN, C. B. (2012) 'Automated intelligent rotor tine cultivation and punch planting to improve the selectivity of mechanical intra-row weed control', *Weed Research*, **52**(4), pp. 327–337.

RASMUSSEN, J. AND SVENNINGSEN, T. (1995) 'Selective weed harrowing in cereals', *Biological Agriculture and Horticulture*. Taylor & Francis Group, **12**(1), pp. 29–46.

REDDY, K. N. AND NORSWORTHY, J. K. (2010) 'Glyphosate-Resistant Crop Production Systems: Impact on Weed Species Shifts', in *Glyphosate Resistance in Crops and Weeds: History, Development, and Management*, pp. 165–184.

REID, J. F. AND SEARCY, S. W. (1988) 'An Algorithm for Separating Guidance Information from Row Crop Images', *Transactions of the ASAE*, **31**(6), pp. 1624–1632.

REISER, D., MIGUEL, G., ARELLANO, M. V., GRIEPENTROG, H. W. AND PARAFOROS, D. S. (2016) 'Crop row detection in maize for developing navigation algorithms under changing plant growth stages', in *Advances in Intelligent Systems and Computing*, pp. 371–382.

RICHTER, O., ZWERGER, P. AND BÖTTCHER, U. (2002) 'Modelling spatio-temporal dynamics of herbicide resistance', *Weed Research*. John Wiley & Sons, Ltd (10.1111), **42**(1), pp. 52–64.

RIEMENS, M. M., VAN DER WEIDE, R. Y., BLEEKER, P. O. AND LOTZ, L. A. P. (2007) 'Effect of stale seedbed preparations and subsequent weed control in lettuce (cv. Iceboll) on weed densities', *Weed Research*. John Wiley & Sons, Ltd (10.1111), **47**(2), pp. 149–156.

RÖÖS, E., MIE, A., WIVSTAD, M., SALOMON, E., JOHANSSON, B., GUNNARSSON, S., WALLENBECK, A., HOFFMANN, R., NILSSON, U., SUNDBERG, C. AND WATSON, C. A. (2018) 'Risks and opportunities of increasing yields in organic farming. A review', *Agronomy for Sustainable Development*. Springer Paris, p. 14.

RUEDA-AYALA, V., PETEINATOS, G., GERHARDS, R. AND ANDÚJAR, D. (2015) 'A non-chemical system for online weed control', *Sensors*, **15**(4), pp. 7691–7707.

SAVARY, S., FICKE, A., AUBERTOT, J. N. AND HOLLIER, C. (2012) 'Crop losses due to diseases and their implications for global food production losses and food security', *Food Security*. Springer Netherlands, pp. 519–537.

SCHMITZ, J., SCHÄFER, K. AND BRÜHL, C. A. (2014) 'Agrochemicals in field margins-Field evaluation of plant reproduction effects', *Agriculture, Ecosystems and Environment*. Elsevier, **189**, pp. 82–91.

SCOTT, R. K., WILCOCKSON, S. J., & MOISEY, F. R. (1979). The effects of time of weed removal on growth and yield of sugar beet. *The Journal of Agricultural Science*, **93**(3), 693-709.

SLAUGHTER, D., CHEN, P., AGRICULTURE, R. C.-P. AND 1999, U. (1999) 'Vision Guided Precision Cultivation', *Precision Agriculture*, **1**(2), pp. 199–216.

SLAUGHTER, D., GILES, D., IN, D. D.-C. AND ELECTRONICS AND 2008, U. (2008) 'Autonomous robotic weed control systems: A review', *Computers and Electronics in Agriculture*, **61**(1), pp. 63–78.

SMITH, R. G., RYAN, M. R., MENALLED, F. D., HATFIELD, J. L. AND SAUER, T. J. (2011) 'Direct and Indirect Impacts of Weed Management Practices on Soil Quality', in *researchgate.net*.

SWANTON, C. J. AND WEISE, S. F. (1991) 'Integrated Weed Management: The Rationale and Approach', *Weed Technology*. Cambridge University Press, **5**(3), pp. 657–663.

TAKAI, R., YANG, L. AND NOGUCHI, N. (2014) 'Development of a crawler-type robot tractor using RTK-GPS and IMU', *Engineering in Agriculture, Environment and Food*. Elsevier, **7**(4), pp. 143–147.

TEASDALE, J. R. (1996) 'Contribution of cover crops to weed management in sustainable agricultural systems', in *Journal of Production Agriculture*. American Society of Agronomy, Crop Science Society of America, Soil Science Society of America, pp. 475–479.

TEASDALE, J. R. AND FRANK, J. R. (1983) 'Effect of Row Spacing on Weed Competition with Snap Beans (Phaseolus vulgaris)', *Weed Science*. Cambridge University Press, **31**(1), pp. 81–85.

TILLETT, N., HAGUE, T., GARFORD, P. AND WATTS, P. (2003) *Development of a commercial vision guided inter-row hoe: achievements, problems and future directions, Precision Agriculture*.

TILLETT, N., HAGUE, T., GRUNDY, A., ENGINEERING, A. D.-B. AND 2008, U. (2008) 'Mechanical within-row weed control for transplanted crops using computer vision', *Biosystems Engineering*, **99**(2), pp. 171–178.

VANGESSEL, M. J., SCHWEIZER, E. E., WILSON, R. G., WILES, L. J. AND WESTRA, P. (1998) 'Impact of Timing and Frequency of In-Row Cultivation for Weed Control in Dry Bean (Phaseolus vulgaris)', *Weed Technology*. Cambridge University Press, **12**(3), pp. 548–553.

VASILEIADIS, V. P., VAN DIJK, W., VERSCHWELE, A., HOLB, I. J., VÁMOS, A., UREK, G., LESKOVŠEK, R., FURLAN, L. AND SATTIN, M. (2016) 'Farm-scale evaluation of herbicide band application integrated with inter-row mechanical weeding for maize production in four European regions', *Weed Research*. Edited by M. Liebman. John Wiley & Sons, Ltd (10.1111), **56**(4), pp. 313–322.

VINCENT, C., PANNETON, B. AND FLEURAT-LESSARD, F. (2013) *Physical Control Methods in Plant Protection*. Edited by C. Vincent, B. Panneton, and F. Fleurat-Lessard. Berlin, Heidelberg: Springer Berlin Heidelberg.

VAN DER WEIDE, R. Y., BLEEKER, P. O., ACHTEN, V. T. J. M., LOTZ, L. A. P., FOGELBERG, F. AND MELANDER, B. (2008) 'Innovation in mechanical weed control in crop rows', *Weed Research*, pp. 215–224.

WILES, L. AND BRODAHL, M. (2004) 'Exploratory data analysis to identify factors influencing spatial distributions of weed seed banks', *Weed Science*. Cambridge University Press, **52**(6), pp. 936–947.

WILTSHIRE, J. J. J. J., TILLETT, N. D. AND HAGUE, T. (2003) 'Agronomic evaluation of precise mechanical hoeing and chemical weed control in sugar beet', *Weed Research*. Blackwell Science Ltd, **43**(4), pp. 236–244.

WISLER, G. C. AND NORRIS, R. F. (2005) 'Interactions between weeds and cultivated plants as related to management of plant pathogens', *Weed Science*, **53**(6), pp. 914–917.

ZIMDAHL, R. (2004) *Weed-crop competition: A Review*. 2nd editio. Ames: Blackwell Publishing Ltd.

ZIMDAHL, R. L. (2015) *Six Chemicals That Changed Agriculture*, *Six Chemicals That Changed Agriculture*.

VAN ZUYDAM, R. P., SONNEVELD, C. AND NABER, H. (1995) 'Weed control in sugar beet by precision guided implements', *Crop Protection*, **14**(4), pp. 335–340.

7 EIDESSTATTLICHE VERSICHERUNG

Eidesstattliche Versicherung über die eigenständig erbrachte Leistung

gemäß § 18 Absatz 3 Satz 5 der Promotionsordnung der Universität Hohenheim für die Fakultäten Agrar-, Natur- sowie Wirtschafts- und Sozialwissenschaften

1. Bei der eingereichten Dissertation zum Thema

"Evaluation of sensor-based precision methods for mechanical weed control in arable crops"

handelt es sich um meine eigenständig erbrachte Leistung.

2. Ich habe nur die angegebenen Quellen und Hilfsmittel benutzt und mich keiner unzulässigen Hilfe Dritter bedient. Insbesondere habe ich wörtlich oder sinngemäß aus anderen Werken übernommene Inhalte als solche kenntlich gemacht.

3. Ich habe nicht die Hilfe einer kommerziellen Promotionsvermittlung oder -beratung in Anspruch genommen.

4. Die Bedeutung der eidesstattlichen Versicherung und der strafrechtlichen Folgen einer unrichtigen oder unvollständigen eidesstattlichen Versicherung sind mir bekannt. Die Richtigkeit der vorstehenden Erklärung bestätige ich. Ich versichere an Eides Statt, dass ich nach bestem Wissen die reine Wahrheit erklärt und nichts verschwiegen habe.

8 DANKSAGUNG

Ich bedanke mich bei Prof. Dr. Roland Gerhards für die Möglichkeit der Promotion und die Betreuung als Doktorand in den vergangenen Jahren. Meinen Arbeitskolleginnen und Arbeitskollegen aus dem Fachgebiet der Herbologie möchte ich für die ausgesprochen gute Zusammenarbeit, entspannte Arbeitsatmosphäre und die gemeinsame Zeit danken. Auch danken möchte ich den Mitarbeitern der Versuchsstation Ihinger Hof für die Hilfe bei der Durchführung meiner Versuche.

Ein besonderer Dank gilt meinen Eltern für das Ermöglichen meines Studiums, den ständigen Rückhalt in schwierigen Zeiten und die kontinuierliche Unterstützung bei der Verfolgung meiner Ziele.

Sehr dankbar bin ich ebenfalls Markus Endreß, der mir schon im Studium als guter Freund mit Rat und Tat zur Seite stand und mich auch während meiner Promotion unterstützt und ermutigt hat.

9 CURRICULUM VITAE

Personal Data

Name	Jannis Machleb
Date and Place of Birth	20.02.1990, Hildburghausen, Germany

University Education

04/2016 – 08/2020	Doctorate candidate of the department of Weed Science, Institute of Phytomedicine, University of Hohenheim
10/2013 – 02/2016	Studies in Agricultural Sciences, University of Hohenheim Master of Science (M.Sc.)
10/2012 – 09/2013	Studies in Agribusiness, University of Hohenheim
10/2009 – 12/2012	Studies in Agricultural Sciences, University of Hohenheim Bachelor of Science (B.Sc.)

School Education

09/2006 – 07/2008	High School, International Baccalaureate Diploma, Seoul Foreign School, Seoul, South-Korea
09/2000 – 08/2006	Secondary School, Deutsche Schule Seoul, Seoul, South-Korea

www.ingramcontent.com/pod-product-compliance
Ingram Content Group UK Ltd.
Pitfield, Milton Keynes, MK11 3LW, UK
UKHW022000190726
13853UKWH00004B/1639

9 783736 972483